ASTRONOMY

Fifth Edition

E. G. Ebbighausen

Late of the University of Oregon

Charles E. Merrill Publishing Company
A Bell & Howell Company
Columbus   Toronto   London   Sydney

MERRILL EARTH SCIENCE SERIES
ROBERT J. FOSTER, EDITOR

Published by
Charles E. Merrill Publishing Company
A Bell & Howell Company
Columbus, Ohio 43216

This book was set in Frutiger.
Text Design: Cynthia Brunk
Production Coordination: Rebecca Money Bobb
Cover Design: Cathy Watterson
Cover Photo: Halley's Comet, © 1978 Lowell Observatory and NOAO/Kitt Peak

International Standard Book Number: 0-675-20413-5
Library of Congress Catalog Card Number: 84-61732
    2 3 4 5 6 7 8 9—90 89 88 87 86 85
Printed in the United States of America

in memoriam

In July of 1984, as this book was being produced, Dr. Edwin G. Ebbighausen passed away. This fifth edition of *Astronomy* is dedicated to his memory in recognition of his contributions to the science of astronomy.

Preface

For thousands of years astronomy has been a most provocative physical science. Our sense of wonder about the motions, distances, and physical nature of the heavenly bodies and the universe has continued to develop as we discover more. Interest in astronomy received its greatest modern impetus in 1610 when, for the first time, Galileo used a telescope to study the sky. Since then the flood of discoveries has reached the point where many scientists regard astronomy as the most challenging of the physical sciences. There are good reasons for calling astronomy the "endless frontier."

The rapid development of astronomy since 1610 has resulted in part from the building of larger telescopes to gather more light and the continuous effort to develop new techniques for light analysis. Because our only samples of the universe come from earth and moon rocks and meteorites, the study of light radiation remains our only method of determining for celestial objects such properties as temperature, size, density, mass, and chemical composition. By now we have devised detection methods to study the available radiation spectra from the shortest wavelength X rays through the ultraviolet, visible, and infrared regions and on into the long-wave radio regions.

However, mere data acquisition is not enough. One must know how to interpret the data. This task requires the discovery of the physical laws, which govern the behavior of matter and are assumed to hold true everywhere in the universe. Some of these laws concern gravity, the nature of electrical and magnetic forces, and the behavior of gases and nuclear forces. Astronomers have applied them with great success. These laws have also challenged experimental and theoretical physicists with the discovery of matter under conditions of temperature, density, and mass that far exceed those found on the earth. The result is a constant interchange of data and ideas between astronomers and physicists and an influx of physicists into the astronomical field.

This short text covers for the nonprofessional reader a rather wide range of astronomical topics with the use of no more than the most elementary algebra.

The text goes beyond the recital of facts and attempts to give the student a feeling for the basic physical behavior of the objects in the universe and the universe as a whole.

Chapter 1 is a brief history of astronomy through the discovery of the law of gravity. This material shows that ideas are constantly changing, and this theme pervades the whole text. The student may discover that the history of science is really a study of a continuous, quiet, but vigorous revolution in changing concepts.

Chapter 2 covers the proofs of the earth's rotation and its revolution about the sun, the seasons, and the water tides on the earth. Much of the chapter is devoted to observations of the moon from the earth and many up-to-date details of the recent results of lunar exploration by space vehicles. The chapter concludes with a concise discussion of eclipses of the sun and moon.

Chapter 3, on the tools and methods of the astronomer, covers a wide variety of methods and concepts that the astronomer uses in analyzing the radiation from celestial bodies in order to discover their physical properties. Included are discussions on radiation and its properties, how the astronomer analyzes a spectrum of the radiation, and the types of telescopes.

Chapter 4 covers in brief the planets, the now extended information on their satellites, and the asteroids. Included are the latest data obtained by the *Mariner* flights to Mars, Venus, and Mercury and the Pioneer flights to Jupiter. The new discoveries about the satellites of Jupiter and Saturn are remarkable.

Chapter 5 overviews comets, and discusses meteoroids, meteors, and meteorites. This chapter closes with a section on the origin of the solar system.

Chapter 6 begins by detailing solar phenomena. The chapter continues with an introduction to the stars: the constellations, stellar distances and motions, the brightness scale for stars, the types of stellar spectra that are observed, and the information that can be obtained from these spectra.

Chapter 7 outlines, briefly but thoroughly, the tremendous variety of star systems and types. Here are included double stars, star clusters, and stars of variable brightness. An important part relates to our galaxy of stars, its structure, the evidence of gas and dust between the stars, stellar formation, and stellar energy sources. Another interesting part discusses the formation and unusual properties of the highly condensed objects called white dwarfs, neutron stars, pulsars, "black holes," and X-ray stars. Some of these objects are very recent discoveries.

The last chapter introduces the student to those massive star clusters called galaxies, their types and properties, a discussion of the expanding universe of galaxies and its interpretation, and the puzzles in the understanding of extragalactic radio sources.

In conclusion, this short text ranges over a wide variety of ancient and modern astronomical topics. While much of the material is descriptive, the main goal of the text is to allow the student to appreciate how much astronomers have found to tell us.

Contents

3

**THE TOOLS AND METHODS
OF THE ASTRONOMER 37**

4

**THE FIRST SOLAR FAMILY: PLANETS,
SATELLITES, AND ASTEROIDS 57**

5   THE SECOND SOLAR FAMILY: COMETS AND METEOROIDS   85

6   THE SUN: AN INTRODUCTION TO THE STARS   97

7

STELLAR SYSTEMS AND VARIETY AMONG STARS   115

8

GALAXIES AND COSMOLOGY   137

APPENDIX 1

1 The Beginnings of Astronomy—Ptolemy to Newton

The fascination most people feel for astronomy arises from the very special nature of astronomy's concern with the remote and inaccessible regions of the universe. Chemists, physicists, biologists, or archeologists can actually handle and study—in the field or the laboratory—those samples of nature that are of interest to them. But astronomers can do this only when they have in hand a fragment of meteoric material. For the rest, they must study the light of the celestial objects and, with the aid of rather special techniques, attempt to derive useful information and ideas. The average person suspects a touch of magic in the astronomer's work; he may even think that astronomers must be smarter than other scientists—this is probably not true, but one can make a good case for their being the most ingenious. If so, it is because they must be. Their accomplishments are the result of frequent borrowing from the mathematicians, physicists, chemists, geologists, and engineers. But the astronomers have given in return.

The most absorbing discoveries in astronomy are relatively recent. Our knowledge of astronomy in Christ's day was not much less than in A.D. 1600, but in that year it was so vastly less than it is now that we should examine the reasons. In part, the enormous progress of the last three and one-half centuries is due to instruments such as the telescope, first used by Galileo in 1610. His small two-inch spyglass may be contrasted with the immense Russian 236-inch reflector in the Caucasus Mountains. To a large extent, this advance was an engineering achievement. Progress in mathematics was another great boost. This began with the development of calculus in the seventeenth century by Newton and Leibnitz and its application to all branches of the physical sciences. But the greatest progress was the result of a change in attitude and point of view regarding the idea that human beings must be at the center of their universal environment and that their position in the universe is a very special one. Real advancement began about A.D. 1600, when people became willing to consider the revolutionary concept that they were not at the center of the universe.

In the main, this chapter covers the period from about A.D. 150, when Ptolemy wrote his famous book on the geocentric system, to the seventeenth century, with Isaac Newton's mathematical formulation of the law of universal gravitation.

1

1.1 THE EARLIEST ASTRONOMICAL OBSERVATIONS

From the earliest times, people observed the celestial bodies. From the first, they observed the rising and setting of the sun, moon, and stars, and the phases of the moon. In time, they also observed seasonal changes in the sun's noon altitude and the points on the horizon where the sun rises. Recognition of **fixed stars** and the division of stars into constellations must have happened early. Along with the latter came the observation that five of these stars were not fixed, but changed their positions in a mysterious way. These stars we now call the **planets** Mercury, Venus, Mars, Jupiter, and Saturn. All of these planets have a generally eastward motion with respect to the other stars, but sometimes they appear to slide back in the westward direction for a while and then resume the eastward motion. We now call the eastward motion **direct** and the westward motion **retrograde**. The idea that our earth is a planet is of quite recent origin.

At some stage, people must have wondered what structure of the universe could explain all these motions. What kind of mechanical model could be devised to account for all the celestial phenomena and also to predict the future positions of these objects?

1.2 THE GEOCENTRIC SYSTEM—PTOLEMY

Over the centuries a system emerged that was highly successful in accounting for these observed motions. The keystone of this theory was the belief that the earth was the center of the universe, at rest with no motion through space or rotation on an axis. This lack of motion was accepted because no motion could be felt. The idea of uniform motion of translation or rotation could never be experienced in practice, and it would not be understood until Galileo's pioneering work on force and motion and Newton's later formulation of the laws of motion. Even more simply, it seemed obvious that motion through space would result in a constant wind from the direction in which the earth was traveling. Of course, no such wind was observed.

The geocentric system is often called the *Ptolemaic system* because of the work of Claudius Ptolemy in about A.D. 150. His main contribution was a book called the *Almagest* ("the great work") in which he summarized in detail the then-current ideas on the geocentric system. For more than a thousand years after Ptolemy, this system continued to be altered and elaborated. One cannot speak of *the* Ptolemaic system, because it underwent so many revisions. However, it is possible to explain its fundamental features. This we will do in the following sections.

1.3 THE INNER PLANETS—MERCURY AND VENUS

Mercury and Venus are morning and evening stars. Venus is seen more often because it is brighter and is usually seen farther from the sun than Mercury. When one of the planets is east of the sun, it sets after the sun and is an **evening** star. When it is west of the sun, it rises before the sun and is a **morning** star.

According to the geocentric view, both planets were considered to revolve on circles with different centers which were always on a line joining the earth and the sun as in Figure 1.1.

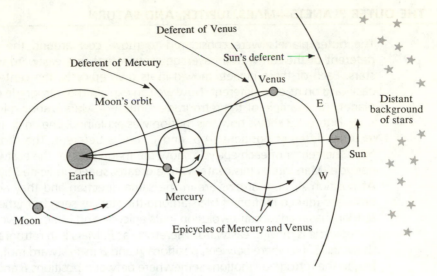

FIGURE 1.1 The motions of Mercury, Venus, the moon, and the sun in the geocentric system. The view is from above the earth's North Pole.

The separate circles for Mercury and Venus are called **epicycles**. Looking from the earth toward the sun in the figure, Venus is seen to the left (east) and is an evening star. In the same figure, Mercury is to the right (west) of the sun and is a morning star. As each planet revolves in its own epicycle in the counterclockwise direction, it becomes alternately a morning and an evening star. At any time, the angle planet-earth-sun is called the **elongation**. This angle at its largest is greatest for Venus, and therefore Venus can be seen farther from the sun than Mercury. It is clear that at midnight neither planet can be seen opposite the earth from the sun.

By the rules of the Ptolemaic system all epicycles are circles. But because the eastward motion of the sun among the stars was known to be nonuniform it was necessary for the sun to move in a small epicycle with a period of one year. Because the eastern and western elongations of Mercury can range from 18° to 28° it would be necessary to have Mercury move on at least one more epicycle whose center moves on the epicycle shown in Figure 1.1. And because the moon's eastward motion is not uniform the moon would need to move on an epicycle whose center moved on the moon's orbit.

1.4 THE MOTION OF THE MOON AND THE SUN

In the geocentric system, the moon revolved about the earth, and the explanation of its phases was the same as the modern one in the heliocentric system (see Section 2.9 in Chapter 2). The sun's motion in Figure 1.1 is counterclockwise (ccw) about the earth in circular orbit called a **deferent**. As observed from the earth, the sun is seen against a distant background of stars. In the geocentric system, as the sun moved in its deferent, it proceeded eastward among the stars a little less than one degree per day, and it came back to the same place among the stars in one year. The view of the constellations at night changed continuously throughout the seasons.

1.5 THE OUTER PLANETS—MARS, JUPITER, AND SATURN

The outer planets were considered to move ccw around the earth beyond the deferent of the sun with an average motion that was eastward with respect to the stars. Each of these planets moved in its own epicycle, the center of which moved eastward on its own deferent. Now we can explain the retrograde motion of an outer planet. An example of the retrograde motion of Mars is shown in Figure 1.2.

Figure 1.3 shows how this motion was explained. The epicycle of Mars is shown in three different positions (A,B, and C) on its deferent. The shorter arrows drawn from the center of each epicycle show the magnitude of the eastward motion of the epicycle. Mars circles this center with a greater speed, indicated by the longer arrow. At position A, both arrows are in the same direction and they add, resulting in an eastward (direct) motion of Mars among the stars as seen from the earth. At position B, Mars has made a half-revolution in its epicycle and is moving westward faster than the epicycle is moving eastward. Therefore, at B, Mars is in retrograde motion among the stars. Somewhere between positions A and B its eastward motion stopped, and it began the retrograde motion. Somewhere between positions B and C, Mars will have stopped its retrograde motion and will have begun to move east again. At position C, Mars will have made a complete revolution in its epicycle and both arrows point directly eastward. Once again, as at position A, the planet will be moving rapidly eastward (direct). The same explanation applies to both Jupiter and Saturn, except that their deferents are larger than that of Mars, and the amplitudes of their retrograde motions are smaller as seen from the earth.

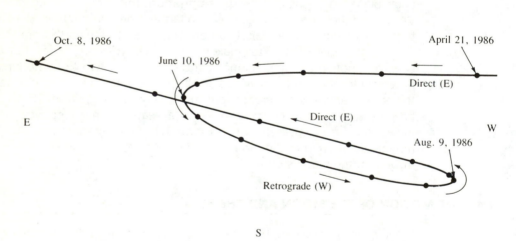

FIGURE 1.2 An example of the retrograde motion of Mars. The dots on the smoothed curve are positions of the planet on the celestial sphere in the southern constellation Sagittarius at ten-day intervals from April 21, 1986, through October 8, 1986. The total amount of E-W retrograde motion is about 12.5 degrees or nearly twenty-five times our own moon's angular diameter. (Data courtesy of the U.S. Naval Observatory.)

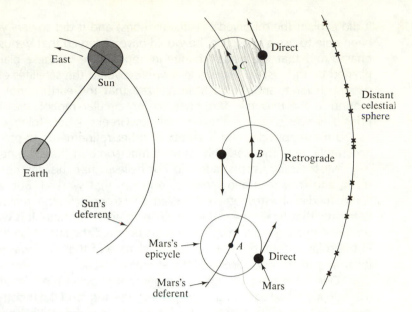

FIGURE 1.3 The epicycle and deferent for Mars to explain the retrograde motion of an exterior planet.

1.6 RISING AND SETTING

In the geocentric system, rising and setting must be explained without the rotation of the earth. Therefore, all the motions described must be accompanied by a westward (cw) rotation of the whole sky in a period of one day. Imagine a large circular platform with a central spot where you (representing the earth) stand with your feet on the ground. Around you on the platform move the moon, sun, and the planets on circular tracks (the deferents), with you as the center. Around the edge of the circular platform, and attached to it, is a large canvas backdrop studded with stars. The average motion of all the objects with respect to the stars will be to the left (eastward). Standing still and facing in one direction, let the whole platform rotate from left to right and make one revolution in one day. Objects will appear from behind you on your left, pass in front, and disappear behind you on your right. Thus rising and setting are explained for a stationary earth. The entire sky and all objects on it move all the way around and back in one day.

1.7 COMMENTS ON THE PTOLEMAIC SYSTEM

Modern readers are well aware that the earth rotates on its axis and revolves in an orbit around the sun. We have presented these sections to show that for thousands of years people thought otherwise. How was it possible to be so wrong for so long a time? For one thing, there was no evidence that the geocentric system was wrong. Based on the principle that the earth was at rest and centered in the universe, this system was a permissible one. Just as important was the fact that the system *worked*.

It did explain the observed celestial motions, and it did so very well. If Ptolemy had been able to use a telescope, he would have observed that Venus and Mercury have phases and that the actual change in appearance of these planets cannot be explained by this system. He would have discovered the satellites of Jupiter, as Galileo did much later, and would have realized that the earth is not the only center of motion in the universe. The insistence on circular motion was based on aesthetic principles, which were not taken lightly by Greeks. After Ptolemy, later modifications added more epicycles and a variety of other refinements to predict positions more accurately, until the system became cumbersome in the extreme.

Most Greek astronomers did not believe that this description was the actual truth, and they regarded it only as a model that worked. But, long after Ptolemy, some medieval astronomers believed so strongly in the reality of epicycles and deferents that finally the system became an article of faith. It is well to realize that in the pre-Christian era a few astronomers believed the sun to be the center of motion in the solar system, but the intellectual climate of that time was not ripe for such an idea, and their views received little attention.

One interesting version of the geocentric system, as described by Pythagoras (ca. 500 B.C.), had each of the five planets, the sun, and the moon carried on separate crystal (transparent) spheres, each centered on the earth and distinct from the outermost sphere of the stars. The slight frictional motion of these spheres moving within each other was said to give rise to the "music of the spheres," a celestial concert which could only be heard by the most perceptive ear.

The view that the Greeks were pure thinkers and not given to measurement is not correct. Actually, a great many celestial measurements were made in the Greek era. For an interesting discussion of this work, the reader is referred to Abell's *The Realm of the Universe,* which is listed in the appendix of suggested readings at the end of this text.

1.8 THE HELIOCENTRIC, OR COPERNICAN, SYSTEM

This system derives its name from Nicholas Copernicus (1473–1543), who was born in what is now Poland. In the year of his death, Copernicus published a book which critically examined and compared the geocentric and heliocentric systems. He placed the sun in the center of the solar system, with all the planets revolving about it. As in the geocentric version, he also regarded the moon as the earth's satellite.

He still believed that the orbits were circles, and he used epicycles to explain nonuniform orbital motion. In the case of Mars, Copernicus showed that, as seen from the sun, the planet moved with a variable orbital speed. To explain this, he had Mars moving in a circular epicycle with a center that moved on a circular orbit (deferent) **centered on the sun.** This is somewhat like the situation discussed in Figure 1.2 and Section 1.5. In this case, the speed in the epicycle was less than the speed of forward motion of the epicycle's center on the deferent. Therefore, as Mars revolved in the epicycle, its forward motion with respect to the stars and as seen from the sun was variable. The retrograde motion of Mars and the exterior planets will be discussed in the next section.

Copernicus did not prove that the sun was the center of the solar system, but he did show that the heliocentric system was far less complicated than the geocentric system.

1.9 RETROGRADE MOTION, COPERNICAN STYLE

In this system, the retrograde motion of an outer planet is explained as the result of the relative motion of the earth and the outer planet. In Figure 1.4, the orbits of the earth and Mars are circles centered on the sun. For simplicity, the epicycle of Mars, as discussed in Section 1.8, is omitted. The numbered points on each orbit are equidistant in time and on the orbit. When earth is at E_1, Mars is at M_1. The spacing of the points on the orbit of Mars is smaller because of its lower orbital speed. If we connect corresponding points on both orbits and continue the line to the star background, we see that Mars moves east (direct) for a while, stops, and then moves west (retrograde) for a time. Mars then stops again and begins its normal eastward motion once more. No epicycles are needed to explain the retrograde motion. The simplicity of this explanation is one of the strongest arguments for the heliocentric system. The same applies to the motion of Jupiter and Saturn, except their orbits are larger and their retrograde motions smaller. The motions of Mercury and Venus are simply accounted for by motion in orbits inside that of the earth and revolving about the sun. In order to explain the nonuniformity of the orbital motion of all the planets, Copernicus used epicycles as in the geocentric system. However, as we shall see shortly, Kepler made the use of epicycles unnecessary.

Mercury and Venus revolve about the sun in the same direction as the earth, and as seen from the earth, each planet independently moves from the east side of the sun to the west side in retrograde motion and then back to the east side in direct motion. This motion is centered on the sun. However, because the sun always moves eastward among the stars as a result of the earth's orbital motion, the sun carries

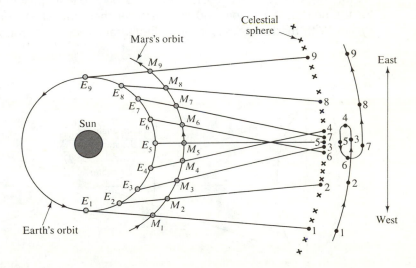

FIGURE 1.4 The retrograde motion of Mars on the heliocentric system.

these two planets with it, so that a complete path from direct to retrograde motion and back again moves continuously eastward against the stellar background as seen from the earth.

1.10 TYCHO BRAHE (1546–1601)

Tycho was one of the most remarkable of all astronomers. Born in Sweden, he was later able to persuade the Danish king to finance an observatory on the island of Hven near Copenhagen. For the determination of planetary and stellar positions he built and used a number of excellent instruments with a previously unexcelled accuracy of one minute of arc. During his lifetime he and his wife, who was also an astronomer, recorded many thousands of positions, particularly of Mars.

One of his most important contributions was to devise and perform an experiment to distinguish between the correctness of the geocentric and heliocentric systems. In the latter, the earth is in orbital motion about the sun. This should result in an annual displacement of nearby stars with respect to the background of more distant stars, as shown in Figure 1.5. As seen from the sun, the star will remain fixed during the year, but from the earth, shown in two positions six months apart at E_1 and E_2, the star will have shifted its position by angle E_1–star–E_2. The more distant the star, the smaller will be the angle. Half of this angle, or the angle E–star–sun, is the **stellar parallax.** If the direction of the star is perpendicular to the plane of the earth's orbit, the parallactic motion will be a circle with respect to the distant star background. If the star is in the plane of the earth's orbit, the shift will be in a straight line. For intermediate positions, the motion will be in the form of an ellipse.

Tycho made this test for a few bright stars which he assumed to be close, but he obtained a negative result. Not until 1838 was this done with a positive result, using much more accurate instruments. Tycho failed because his measurements were too crude. For the closest known star the parallax is about three-quarters of a second of arc (0."76). One second of arc is 1/3600 of a degree in angular measure. The parallax for the nearest star is some 80 times smaller than any which could be observed by Tycho. With modern telescopes and photographic plates astronomers can measure

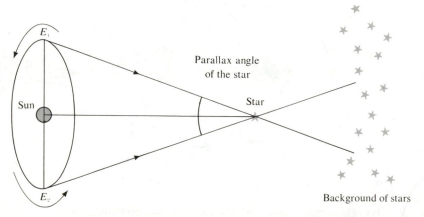

FIGURE 1.5 Tycho's test for the annual parallax of a star.

THE BEGINNINGS OF ASTRONOMY–PTOLEMY TO NEWTON

with confidence a parallax of 0".02, 2/100 of a second of arc. The parallax angle and the distance are related by the formula $p'' = 1/D$, where D is the distance at which the parallax would be one second of arc. This distance has the name **parsec**, and it is the technical unit of stellar distance commonly used by astronomers. One parsec equals 206,265 astronomical units, or 3.26 light years. As we shall see later, the **astronomical unit** (AU) is the average distance of the earth from the sun and is the standard of distances in the solar system. The **light year** is the popular unit of distance and is the distance that a ray of light will travel in one year. Light moves in a vacuum at the rate of approximately 300,000 kilometers per second, or 186,000 miles per second. One light year is very nearly 6 trillion miles. The conversion of English units to those in the metric system is given in Appendix 2.

1.11 JOHANNES KEPLER (1571–1630)

After falling out of favor with the Danish king, Tycho went to Prague, Czechoslovakia, in 1597. In the year before Tycho's death, Johannes Kepler, a distinguished young mathematician, became his assistant. After Tycho's death, Kepler acquired the immense body of planetary observations made by Tycho and his wife and began a thorough investigation of planetary motions. He spent most of a 25-year period on the problem and achieved success because he was willing to break with tradition and to discard deferents, epicycles, and uniform motion. It may be said that Kepler's work marks the beginning of modern astronomy. His famous three laws of planetary motion will now be discussed.

1.12 KEPLER'S FIRST LAW: THE FORM OF THE ORBIT

Kepler's first law states: "The form of the orbit of a planet is an ellipse with the sun at one of the foci." To construct an ellipse, place a sheet of paper on a board and drive into the board two nails a short distance apart, as in Figure 1.6(a). Form a knotted string with a total length of a little more (say, 20 percent) than twice the distance between the two nails. Place the loop of string over the nails and draw it tight, using a pencil tip. Move the pencil all the way around the nails, keeping the loop tight. The resultant curve will be an ellipse and the position of each nail will be one of the **foci** (singular, **focus**). In the same ellipse in Figure 1.6(b), the diameter through the two

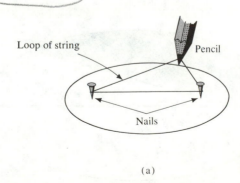

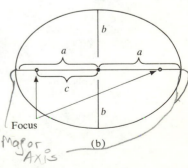

FIGURE 1.6 The ellipse.

foci will be the **major axis,** and from the center of the ellipse to either end will be the **semimajor axis.** The distance (c) divided by the semimajor axis (a) is defined as the **eccentricity** (e) of the ellipse. Note that for the same length of string, the shape of the curve will depend upon the distance between the two nails. When it is zero, the curve will be a circle (constant radius) and the eccentricity will be zero. The semiminor axis b is shown in Figure 1.6(b). When the eccentricity is zero and the ellipse is a circle, a and b are equal. As the eccentricity increases from zero, b decreases with respect to a. In the case of a planetary orbit the sun is at one of the foci—there is nothing at the other one. The point where the planet is closest to the sun is called **perihelion.** The most distant point at the other end of the major axis is called **aphelion.** A little algebraic juggling will show that at perihelion the planet's distance is a (1 − e) and it is a (1 + e) at aphelion.

1.13 KEPLER'S SECOND LAW: THE LAW OF AREAS

Even in the Greek era it was known that the motion of the sun among the stars was not the same at all times of the year and that this difference accounted for the different lengths of the seasons. To take this nonuniformity into account, a small epicycle which moved on the sun's deferent was added. The new view of Kepler was that the nonuniform motion resulted from the fact that the earth's orbit was an ellipse and that its orbital velocity changed through the year. Kepler found that this was particularly true for Mars. This led in time to the second law, which states that the line (the radius vector) joining a planet to the sun sweeps over equal areas in equal intervals of time. This was a brilliant discovery. The meaning of this statement is shown in Figure 1.7.

Here a, b, c, and d are points on the orbit so spaced that the time interval between successive positions is the same, say one year. The second law requires that the area bounded by the arc between two successive points and the two radius vectors from the sun to these two points will be the same for each successive pair of points. This means that when the distance from the sun is relatively large the orbital velocity will be small, so that the area swept out by the radius vector will be the same as when the distance to the sun is less and the orbital velocity must be larger. Thus, the orbital velocity will be largest at perihelion and smallest at aphelion.

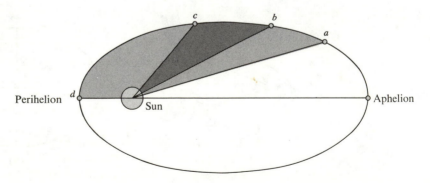

FIGURE 1.7 The law of areas.

With the use of Tycho's observations, Kepler was able to determine for each planet that its orbit is an ellipse. Kepler measured the length of the semimajor axis in terms of the earth's orbit, determined its orbital eccentricity, and also derived the orbital period of the planet around the sun. After a long study of these data, he came to the conclusion that there was a relationship (a "harmony") between the semimajor axes (*a*) of the planets' orbits and their revolution periods around the sun. Kepler showed that if the length of the semimajor axis of the orbit of *any* planet is expressed in astronomical units (AU), and its period of revolution is given in years, then

$$\frac{a^3_{AU}}{P^2_{yr}} = 1 \quad \text{or} \quad a^3_{AU} = P^2_{yr}$$

That is, the ratio of the cube of the semimajor axis in astronomical units to the square of the period in years is equal to one (unity); this is true for any planet. This relation can be used to find the semimajor axis of the orbit of any planet (or any other body revolving about the sun) in astronomical units if its orbital period in years is known. Later, Newton was able to obtain Kepler's three laws from the law of gravity. He showed that a more correct form of the third law is

$$\frac{a^3_{AU}}{P^2_{yr}} = m_1 + m_2$$

where the masses (*m*) of the two bodies are measured in terms of the sun's mass being unity. Newton's form of the third law is true for any two bodies revolving about each other under the influence of the force of gravity. This applies if the two bodies are the sun and a planet, a planet and its satellite, or two stars. This form is nearly the same as that obtained by Kepler because if m_1 is the sun's mass, then m_1 is one solar unit. The mass of the planet (m_2) is very much smaller than unity, so the sum of the masses is very close to unity for all the planets. As an example, Jupiter, the most massive planet, has about 1/1000 of the sun's mass. Therefore, in this case, $m_1 + m_2$ = 1.001, which is very close to unity. Within the limits of the errors of determination of Kepler's data, he could not have observed this difference from unity. The mass sum is even closer to unity for the other planets. This second form of Kepler's third law is one to remember because it will be used in later chapters.

Kepler's third law is very useful in obtaining planetary masses if the planet has at least one observable satellite. Then it is not usually difficult to obtain the semimajor axis of the satellite's orbit in kilometers or miles and also the revolution period in days. Next convert the length of the semimajor axis into astronomical units and the period into years and substitute these values into the second form of Kepler's third law. The result will give the sum of the masses of both the planet and the satellite ($m_p + m_{sat}$) where the unit is the sun's mass. If the satellite is quite small as compared to the planet, its mass is also comparatively small and the above sum will be a good approximation to the planet's mass.

It should also be noted that Kepler's third law made it possible for the first time to determine the dimension of planetary orbits. Consider the law in the form $a^3_{AU} = P^2_{yr}$. Then determine the orbital period in years with whatever accuracy is permissible, substitute this value in the formula and obtain the semimajor axis in astronomical units. Even if the conversion factor to kilometers or miles is not available, at least one has the relative sizes of the planet's orbits.

1.15 COMMENTS ON KEPLER'S LAWS

Kepler discovered these laws by a brilliant analysis of the large body of observations obtained by Tycho. The laws are simple in form, but their importance is very great, since they showed that laws govern planetary motions. The introduction of elliptical orbits was a great innovation and an important departure from the ancient idea that all motion must be in a perfect circle. The second law is a beautiful and simple description of the change in orbital velocity of a planet. Kepler showed that the third law applied not only to the orbits and periods of revolution of the planets, but also to the motion of a satellite about a planet. In spite of this tremendous achievement, there is something lacking in the laws. They *describe* the motions of the planets and satellites, but they offer no clue to any fundamental explanation based on a law of force. Kepler's laws are magnificent fragments of the greater truth to be discovered some years later by Newton.

1.16 GALILEO GALILEI (1564–1642)

Galileo, as he usually is called, was one of the greatest thinkers, originators, and experimentalists of all time. In his lifetime, he advanced the frontiers of science more than any person before him. Galileo's researches in mechanics had a direct bearing on Newton's formulation of the laws of motion.

Galileo is the first person known to have used a telescope for astronomical observations. He used one as early as 1610. With it, he discovered the moons of Jupiter, the craters on the moon, star clusters, the phases of Venus, and the secret of the Milky Way. He was the first person to study sunspots with a telescope. His discovery of Jupiter's moons showed that the earth was not the only center of motion, as required in the geocentric theory. One of Galileo's greatest contributions was his advocacy of the heliocentric system. On this he wrote copiously and with great penetration, and for his views he was resoundingly criticized by the upholders of the older theory, which had been raised to the level of theological dogma. The story of this revolt and its eventual success is one of great interest. The reader is referred to the many books on the subject.

1.17 ISAAC NEWTON (1643–1727)

After what has been said about Galileo, it would be hard to say more of any other person, except Newton. He is best known for his law of universal gravitation, but he also contributed a great deal to the study of light and optics and formulated the laws of mechanics. Inherent in these laws is the clarification of the meaning of uniform motion, force, and acceleration.

Newton's statement of the law of universal gravitation is that the force of gravity between any two bodies in the universe is directly proportional to the product of their masses and inversely proportional to the square of the distance between their centers. This rather formidable statement can be expressed in symbolic form as

$$F = G \times \frac{m_1 \times m_2}{d^2}$$

where F is the force, m_1 and m_2 are the masses, and d is their separation. The constant of proportionality is called G. Once the units of mass, distance, and force are specified, G tells us the magnitude of the force for given values of the masses and distance. Another way of stating the formula is that the force *depends* upon the product of the masses divided by the square of the distance. Note that the equation says nothing about the temperature, color, physical state, or chemical composition of the masses. It is independent of these attributes.

Consider the effect of distance on the force. The fact that d is in the denominator means that the larger the d, the smaller the force. However, since d is squared, it means that the force changes more rapidly than if the first power were used. For example, it says that if we consider the force between the earth and the moon as the standard and then double, triple, or quadruple the distance, the force will become, respectively, ¼, ⅑, or ⅟₁₆ of the original standard force. If we cut the distance between the earth and the moon to ½ or ⅓ of its present value, the force would become, respectively, 4 times or 9 times larger than its present value. Note that the force approaches zero only as d approaches infinity.

The law of gravity is important because it expresses the behavior of the force of gravity between any two bodies in the universe regardless of their separation or masses. There is no reason to believe that the law has ever been any different in past time or that it will be altered in any way in the future. One result of the law was Newton's demonstration that Kepler's three laws are direct consequences of the law of gravity.

Newton's work marks the end of a very long period of confusion and a slow reaching out for an understanding of order and meaning in the universe. After Newton, a host of persons expanded his work and developed that most intricate branch of mathematical physics called celestial mechanics. At long last, it became possible to develop in detail the intricacies of the motions of the many bodies in the solar system and to understand all motions anywhere in the universe as they depend on the force of gravity.

KEY TERMS

fixed stars	epicycle	light year
planets	elongation	Kepler's three laws
direct motion	deferent	foci (focus)
retrograde motion	heliocentric	semimajor axis
geocentric system	Copernican	eccentricity
Ptolemaic system	parallax	perihelion
evening star	parsec	aphelion
morning star	astronomical unit	Isaac Newton

1. How would the changing pattern of the phases of Venus differ between the heliocentric and geocentric explanations of motions in the solar system?

2. Consider the case of a comet moving around the sun in a very highly eccentric orbit with a period of 10,000 years. Approximately how far from the sun will the comet be at aphelion?

3. Using the data on Mars in Table 4.1, compute the perihelion and aphelion distances of that planet.

4. Draw a series of ellipses under the conditions that (a) the length of the loop of string remains the same, but the distance between the nails is changed, and (b) the distance between the nails is fixed, but the length of the loop of string is altered.

5. Summarize the arguments in support of the Ptolemaic and Copernican systems.

6. If the distance between the earth and sun were doubled and the mass of the earth were tripled, how would the force of gravity of the earth on the sun compare with its present value?

7. Discuss the importance of Kepler's three laws of planetary motion.

8. What explanation would you give for the fact that Venus is not seen to pass between the earth and the sun at each inferior conjunction?

9. How can one determine the mass of a planet if it has a satellite?

10. Show that at perihelion Pluto is closer to the sun than is Neptune. See the data in Table 4.1.

2 The Earth and the Moon

Because the earth is rotating, the resulting centrifugal force changes the earth's shape from a sphere to an oblate spheroid with an equatorial diameter of 12,762 km (7927 mi) and a polar diameter of 12,720 km (7900 mi). The mass of the earth is 6.6 $\times$ 10^{21} tons. When this number is divided by the volume of the earth, its average density is found to be 5.5 times that of water. Since the average density of the surface rocks is 2.7 times that of water, it is clear that some parts of earth's inner regions must have a higher density than the average for the whole body. The existing evidence suggests that the density near the center is 12–15 times that of water. The earth is the most massive of the four inner planets. Among these planets, the earth is the only one that has liquid surface water and also a temperature and atmosphere sufficient to permit the existence of life as we know it.

2.1 THE PROOFS OF THE ROTATION AND REVOLUTION OF THE EARTH

As shown in the first chapter, the rotation of the earth explains the rising and setting of the heavenly bodies. However, the proof of its rotation was not obtained until 1851 when Foucault performed his famous experiment with a pendulum in Paris.

To understand his experiment, imagine a free swinging pendulum hung from the ceiling of a high room at the North Pole of the earth. As seen from above, the plane of the swing would gradually turn in a clockwise direction with respect to the room, giving the impression that some force acts on the pendulum to cause its plane of swing to change. However, the only force acting on the pendulum is gravity, which acts vertically *downward* and hence cannot influence the plane of swing. Also, the plane of swing would make one 360-degree clockwise turn in one day. Thus, we would conclude that the plane of swing of the pendulum does *not* change, but rather the room and the earth turn under the pendulum in a counterclockwise direction. That is, the earth is turning from west to east.

If the same arrangement were set up at the equator and the pendulum were set in motion in any direction, one would observe that the plane of swing does not change with time with respect to the walls of the room. That is because at the

15

equator, while the room is moving around with the earth, it is not turning around a line drawn from the room to the earth's center.

At the South Pole, the effect would be the reverse of that at the North Pole, and the plane of swing would move in the counterclockwise direction with respect to the room walls. Hence, the earth is turning on its axis from west to east once per day. In latitudes between the poles and the equator, the plane of swing makes a complete turn in more than one day, increasing from one day at the poles to an infinitely long time at the equator. At Paris, Foucault observed the period to be about 32 hours.

There are several proofs of the earth's revolution about the sun, but only one will be discussed here. This one involves the phenomenon of stellar parallax discussed in the first chapter under Section 1.10 and diagrammed in Figure 1.5. If the earth were at rest in the universe with no orbital motion, then no parallactic motion would be observed for nearby stars with respect to distant ones. However, parallactic motion *is* observed, and the size of the parallax has been measured for a few thousand stars. Equally important, parallactic motion, whether in a circle, an ellipse, or a straight line, goes through one complete oscillation in *one year*, for all stars. This is the period of revolution of the earth about the sun. If the stars could be observed from the sun, we would see no parallactic motion for any star.

2.2   THE ORBIT OF THE EARTH

The earth's orbit has an eccentricity of 0.017 (Section 1.12). The semimajor axis is about 150 million km (92,900,000 mi), with an uncertainty of just a few miles. Because of the eccentricity, the earth is closer to the sun by 1.5 million miles in early January and farther from the sun by the same amount in early July than if the orbit were a circle. Because of the law of areas (Section 1.13), the earth's orbital velocity is 1.7 percent greater than the average (29.8 km/s or 18.5 mi/s) in January and less by the same amount in July.

The path of the sun among the stars as seen from the earth is called the **ecliptic**, and the earth's orbital plane is called the **plane of the ecliptic**. The **zodiac** is that band in the sky 9 degrees on either side of the ecliptic, and in this band are found the sun, the moon, and the planets (with the occasional exception of the moon, Venus, and Pluto). The **signs of the zodiac** are twelve 30-degree divisions along the ecliptic, each having the name of a constellation. These signs are of interest only in astrology.

2.3   POSITIONS ON THE EARTH AND IN THE SKY

Because the earth is in rotation there is an axis which pierces the surface at the **north** and **south terrestrial poles**. That great circle on the earth, all points on which are equidistant between these poles, is the **terrestrial equator**. The **latitude** of a place on the earth's surface is measured in degrees north (N, +) or south (S, −) of the equator. The east-west distance of the place is called **longitude** and is a measure of the angle at either pole between the **prime meridian** and that meridian passing through the place. A meridian is any circle on the earth's surface from the north to the south pole. The prime or standard meridian is by international agreement that which passes through the old Greenwich, England, observatory. The longitude may also be the arc on the equator between the two meridians. This coordinate may be measured in time up to 12 hours or in degrees up to 180 either east or west of the prime meridian.

On looking at the sky, day or night, one has the impression of being on the inside of a giant bowl, the **celestial sphere**. All celestial objects appear to be on that imaginary sphere. For any observer on the earth's surface the **zenith** is that point on the celestial sphere directly above the observer and as defined by the direction of a cord suspending a weight. The **altitude** of an object on the celestial sphere is its angular elevation above the horizon measured on the vertical circle passing through the object. The **azimuth** of the object is the angular measure in degrees along the horizon circle from the north point in the eastward direction to the vertical circle through the object.

When the earth's rotational axis is projected in both directions into space, we can imagine it to intersect the celestial sphere at the **north and south celestial poles** (NCP and SCP). See Figure 2.1. Just as on the earth, the **celestial equator** is that great circle on the sky on which all points are equidistant from the two celestial poles. The extension of the plane of the earth's equator out to the celestial sphere will cut the latter along the celestial equator. To a person standing at the north terrestrial pole, the north celestial pole will be overhead in the zenith, and the celestial equator will lie along the horizon. At the earth's equator, the celestial equator will pass through the observer's zenith, and the two celestial poles will be, respectively, at the north and south horizon points. Here days and nights will be of equal length on all days of the year. At all latitudes, the celestial equator will cut the horizon at the east and west points, except that at the North and South Poles of the earth, the celestial equator will coincide with the horizon. For these two extreme cases and for inter-mediate latitudes, the altitude of the north celestial pole above or below the north horizon is equal to the latitude of the observer. Because of this equality one can determine one's latitude in either hemisphere from altitude observations of a star close to either pole. In the northern hemisphere this would commonly be the bright star Polaris; for the southern hemisphere the star commonly chosen is the consider-ably fainter star (but visible to the naked eye) sigma Octantis. The altitude of either star in its own hemisphere plus tabular data and formulae from the Astronomical Almanac will provide an accurate value of the observer's latitude. This almanac may be obtained from the U.S. Naval Observatory.

Because the earth's axis is tilted by 23.5 degrees with respect to the normal to the plane of its orbit, the ecliptic is tilted by the same angle to the celestial equator, as seen in Figure 2.1. From the figure, one can see that the ecliptic crosses the celestial equator at two points called the equinoxes. The vernal equinox is the point at which the sun, moving from south to north, crosses the celestial equator on about March 21; this marks the beginning of the spring season. On about September 22, at the autumnal equinox, the sun moves from north to south across the celestial equator, marking the beginning of the fall season. On about June 21, the sun is farthest north of the celestial equator at the point called the summer solstice and farthest south on about December 21 at the winter solstice. June 21 and December 21 are the beginnings, respectively, of the summer and winter seasons. At all points on the earth's equator and at all times of the year, the days and the nights are of equal length. The astronomer measures the position of a star north or south of the celestial equator in degrees; this position is called the **declination**. Longitude in the sky is

FIGURE 2.1 The celestial sphere.

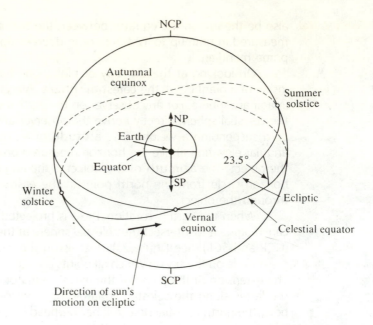

called **right ascension** and is measured in hours eastward through 24 hours from the prime celestial meridian through the vernal equinox.

2.4 THE SEASONS

As stated in the last section, the earth's axis is tilted by 23.5 degrees from the vertical to the plane of its orbit. For the time being we shall assume that as the earth revolves about the sun both the tilt and direction of the axis in space will remain the same. An exception will be noted in the next section. Figure 2.2(a) represents the earth at four positions in its orbit during the year and shows the direction of the sun's rays with respect to that of the earth's rotational axis.

On March 21 the earth's axis is at right angles to the direction of the incoming solar rays and the earth is illuminated from pole to pole. Everywhere days and nights have an equal length of twelve hours and the sun is on the celestial equator. The sun rises directly in the east and sets directly west. The sun's rays strike the earth beyond the north pole until on June 21 they reach the limit of 66.5°N (Arctic Circle) (90° − 23.5°). On that day the whole polar cap north of 66.5° is in sunlight. But south of 66.5°S it is dark for 24 hours. On June 21 in the northern hemisphere the sun is directly overhead on the latitude circle of 23.5° N. called the Tropic of Cancer. See Figure 2.2(b). Between March 21 and June 21 the periods of sunlight (the day) become longer than those of darkness (the night). The maximum difference is the greatest on June 21 after which the difference decreases.

Spring and warmer weather begin before June 21 in the northern hemisphere because after March 21 there are more hours of daylight than of darkness, and the sun's rays on the average become more nearly vertical. A six-inch diameter column of sunlight striking the earth perpendicular to its surface supplies more heat per square inch than one whose angle of incidence from the normal is, let us say, 60°, and the

FIGURE 2.2(a) The seasons.

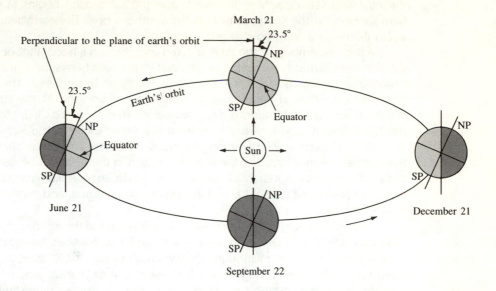

energy is spread over an ellipse six inches wide and twelve inches long. For the normally incident area six inches in diameter, the area is one-half that of the twelve-by six-inch one and hence the heating effect per square inch is twice that of the larger area.

After June 21 the altitude of the sun at noon decreases as seen from middle latitudes in the northern hemisphere, but the reverse takes place in the southern one. By custom summer begins on June 21 in the northern hemisphere, but the warmest part of that season does not take place until at least a month later because of the usual temperature lag. This temperature lag is observed in all the seasons. On September 22, the sun is once again on the celestial equator (autumnal equinox), and

FIGURE 2.2(b) Important latitude circles on the earth.

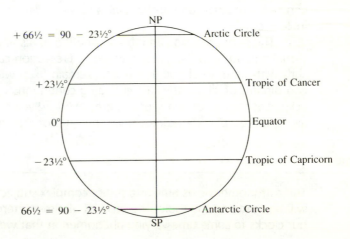

the days and nights are of equal length (12 hours) everywhere, the sun rises directly in the east and sets directly in the west, and the fall season begins in the northern hemisphere and the spring season in the southern one. The earth's surface is once again illuminated from pole to pole.

After September 22 the days shorten and the sun's noon altitude decreases in the northern hemisphere. Spring begins in the southern hemisphere and at the south pole the long antarctic night starts to break. Finally on December 21 the sun is on the Tropic of Capricorn at declination 23.5°S. Again as on June 21 the sun illuminates the polar region for 24 hours a day from 66.5°S (the Antarctic Circle) to the pole. At the same time it is dark all day long from the Arctic Circle to the North Pole.

After December 21 everything happens in reverse. The winter season begins in the northern hemisphere and the summer season in the southern hemisphere. At last the earth reaches the March 21 position. The earth again is illuminated from pole to pole, the periods of daylight and darkness are again equal, and everywhere the sun rises in the east.

It should be clear that these seasons are the result of the inclination of the earth's rotational axis to the plane of its orbit. The same is true for Mars, where the inclination is nearly 24 degrees even though the revolution period is 687 days and the season length is nearly twice as long. A peculiar case is that of Uranus, where the rotational axis lies nearly in its orbit plane, giving rise to some peculiar circumstances.

2.5   THE PRECESSION OF THE EQUINOXES

Although the tilt of the earth's axis remains essentially constant, the direction does not. The direction of the axis changes continually with time in a period of 25,900 years. This is the phenomenon of **precession**. The axis precesses around the normal to the plane of the earth's orbit, just as a spinning top can be seen to do. This precession is the reaction of the earth to the pull of the sun and the moon on the earth's equatorial bulge. At the present time, the earth's axis points in space close to the star Polaris, but it is slowly moving away. About 2700 B.C., the north star was the star Thuban in the constellation Draco. Some 26,000 years from now, the pole star will once again be Polaris. Because of precession, the equinoxes slide westward along the ecliptic continuously. Some 2000 years ago, the vernal equinox was in the constellation Aries, one of the twelve in the zodiac, but now it has moved westward into Pisces.

The right ascension and the declination of a star are defined with respect to the vernal equinox and the celestial equator. Precession causes this coordinate system to shift with time, so although the stars remain fixed with respect to one another, the coordinates of all the stars are changing continuously. The equinoxes slide along the ecliptic at the rate of nearly one degree every 72 years. The vernal equinox has moved along the ecliptic 25 degrees since the time of Ptolemy.

2.6   TIME

The astronomer views time as a rather complex subject, and here only a few remarks will be made on the topic. The basis of all time determinations is the ability to refer our clocks to some time-varying phenomenon that we hope is of constant period for all time. For ages we have used the rotation period of the earth (the day) as our standard and have defined the second as 1/86,400 part of this period. However, for the last hundred years we have known that the earth is gradually slowing down, with the result that the day is lengthening. This is mainly because of the frictional effect of the ebb and flow of the water tides of the earth. This is particularly effective in shallow bodies of water like the Bering Sea. When astronomers compute the times of occurrence of solar eclipses that took place two thousand years ago, they find about a three-hour difference between the observed and computed times.

For most human purposes this longer day matters very little, if at all, but for some astronomical purposes it is most important. The scale of this effect can be understood from the fact that a clock started in 1900 and adjusted continuously to the slowing rate of rotation of the earth would be, in 1986, nearly 56 seconds behind another clock started at the same time and rated to run always at the rotation period of the earth in 1900. At present, astronomers have adopted **ephemeris time**, in which the second is what it was in 1900, as opposed to **universal time**, which uses the instantaneous rate at any time.

It would seem that the period of rotation of the earth with respect to the sun would be the ideal basis for our measurement of time even taking into account the rather small errors discussed in the first two paragraphs of this section. However, the length of this day, the solar day, is quite variable because the sun moves eastward on the ecliptic and also because the earth's orbit has an appreciable eccentricity. For general use the astronomers have developed **mean solar time** in which a fictitious sun moves eastward by a constant amount each day. A sun dial gives a reading of solar time. Usually attached to the base of the sun dial is a plaque giving for many days of the year the correction to convert sun dial (solar) time to mean solar time.

But corrected sun dial time is only local mean solar time and that will depend on the longitude of the astronomer. For convenience the local mean solar time of the meridian of Greenwich is called universal time (UT). It begins with 0^h at Greenwich midnight. Tables of all sorts of phenomena are made for this time so that the occurrence of these events will have a common base which may be converted into the UT of the observer.

Sidereal time is a variety commonly used by astronomers. Consider the passage at the same moment of the fictitious sun and a star directly above or below it across the local meridian, and note the passage of these two objects across that meridian again after one complete rotation of the earth. Because the earth is moving (ccw) in its orbit about the sun and changing its position during this one rotation with respect to the star vastly farther away, the observer on the earth will see the star come to the local meridian first and then will see the fictitious sun. The time interval between the two successive passages of the star across the meridian will be a sidereal day and for the fictitious sun will be a mean solar day. The latter day will be longer by close to 3^m56^s. In the course of one year of close to 365¼ mean solar days this interval will add up exactly to one more sidereal day or close to 366¼ sidereal days.

If we measure the time interval from one phase of the moon to the same phase again, it will be seen that it is close to an average of 29½ days. This is called the synodic period—the phase period. But a measure of the time for the moon to move from one position among the stars to the same position again is close to an average of 27⅓ days. This is the sidereal period; the reason for the difference between these two periods is the same as that for the difference between the solar and sidereal days.

A rotation or revolution period of any object with respect to the sun is always the synodic period; that period with respect to the stars is called sidereal. For example, for Mars the synodic revolution period is 780 days; the sidereal period is 687 days.

2.7   THE INTERIOR OF THE EARTH

The average density of the earth's surface rocks is about 2.7 times that of water, but the average density of the whole earth is 5.5 times that of water. The earth's central density is between 12 and 15 times that of water. Geologists believe that the reason for this is two-fold. First, early in its formation the earth was hot and molten enough for the heaviest minerals to sink to the earth's center, and the lighter ones now form the earth's crust. This process is called **chemical differentiation**. Second, there is a strong pressure increase with depth and this in itself causes a density increase.

Measurements made in oil wells and deep mines show that there is an average temperature increase with depth of about 1°C/30 m (1°F/60 ft). A simple calculation will show that, if this temperature gradient was the same all the way to the center, the central temperature would be about 200,000°C (350,000°F), and the whole earth would be molten and gaseous except for a thin solid outer crust. Clearly this model is unacceptable. From a study of the tidal deformation of the earth caused by the sun and the moon, it can be calculated that on the average the rigidity of the earth is greater than that of steel. The result is that the central temperature of the earth cannot be more than a few thousand degrees and the outflow of heat from the interior must be mainly due to radioactive heating in the outer 32 km (20 mi)—the crust. A study of earthquake waves shows that two types of body waves are transmitted through the earth. The P (primary) waves can travel through solids and liquids and are the first to arrive at a particular seismograph station from the place where the energy was released. The S (secondary) waves can travel only through solids, and, because of their slower speed, arrive after the P waves. When an earthquake occurs, seismographs in a circular zone on the opposite side of earth will detect P waves but not S waves. Outside the zone both P and S waves are observed. This can only mean that there must be a central region of liquid material through which the P waves pass but not the S waves. Outside this zone both P and S waves are observed over the rest of the earth's surface; hence the travel zone must be solid. Extensive observations show that on the outside of the earth there is a thin crust, then a solid shell, the **mantle**, with a thickness of about 2900 km (1800 mi), and last, the liquid core, with a radius of 1300 km (800 mi). Although the mantle is solid and fairly hot, it is not completely rigid; as a result it is now known that there are convection currents in the mantle. These currents reach the surface in the ocean floors, where the crust is thinnest, and cause a horizontal spreading which results in a drift of the floating continents. We can with ease examine the surfaces of Mercury, our moon, and Mars and see no evidence for continental drift. Cloud-covered Venus is not a clear-cut case. But because this drift is clear for our earth and depends to a large extent on internal temperature, one can say that this temperature depends on the mass of the body. Hence, as we shall see later, our low-mass moon probably does not have a large fluid core or a hot plastic mantle like our earth. Even intermediate-mass bodies like Mercury and Mars must have too small a mass to exhibit continental drift and the collision of continental plates. If such phenomena were present even on a much-reduced scale as compared to the earth, the motion would have destroyed most of the impact craters that were created four billion or more years ago.

2.8 THE MOON

The semimajor axis of the moon's orbit around the earth is nearly 385,000 km (239,000 mi), but because of its fairly large eccentricity (0.055), its distance from the earth is quite variable. Because the moon's orbit is greatly affected by the perturbations of the earth and the sun, the eccentricity and the semimajor axis change with time. At its closest point to the earth (**perigee**), the moon may be as close as 221,460 miles away. At its farthest distance from the earth (**apogee**), the moon may be as far as 252,710 miles away. This change produces a very perceptible difference in the moon's angular diameter as seen from the earth. The **angular diameter** of an object is the angle between the two lines drawn from the eye to the opposite sides of the object.

The distance to the moon has been determined by triangulation measurements from two stations on the earth, and the error is about one mile. From this distance and its angular diameter of close to 1/2 degree, its linear diameter is 3476 km (2160 mi). However, a new technique for distance measurement is to attach a laser to a large telescope and to send a highly intense short-time pulse of light to the moon. Some of this light is returned to the telescope by reflection from special mirrors placed on the moon by the Apollo astronauts. The time it takes for the pulse to reach the moon and return to the telescope is measured (about 2.6 seconds), multiplied by the velocity of light, and then divided by two. The distance from the telescope to the reflector on the moon can be determined with an accuracy of about 30.5 cm (1 ft). When these measurements are made for a few decades, we shall learn a great deal about the motion of the moon, and when this is done from two or more continents, it will be possible to check the theory of continental drift. At one time, it was thought that the moon had an ellipsoidal shape with its longest axis pointing toward the earth, but recent measurements made from lunar orbiting satellites show clearly that the moon is nearly spherical. The moon's mass is $\frac{1}{81}$ that of the earth. From this mass and the moon's volume, an average density of 3.3 times that of water is calculated. This is only 60 percent of the earth's average density, which is 5.5 times that of water. The pull of gravity on the moon's surface is only one-sixth that at the earth's surface.

2.9 THE TIDES

The moon and the sun produce the water tides on the earth by their gravitational pull. Each body produces two tidal bulges, one between the earth and the attracting body, and the other on the opposite side of the earth. Each body causes two tidal bulges because the attraction of the body on the water between the earth and the body is greater than the attraction of that body on the earth itself. Also, the attraction of the body on the earth is greater than the attraction on the water on the side of the earth opposite to that of the body. The tidal bulges resulting from the moon's gravitational pull are a little more than twice the height of those produced by the sun.

Let us first consider just the lunar (moon) tides and look at Figure 2.3 First of all, let the moon be fixed in its orbit at M_1. The earth's tidal water is outlined by the solid

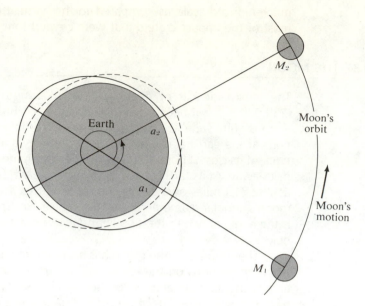

FIGURE 2.3 The earth tides. The view of the earth is from above the North Pole.

elliptical curve, with one bulge toward the moon and the other on the opposite side of the earth. The tide is **high** at point a_1 on the earth and also on the opposite side of the earth. Halfway between these two points, the water is shallow and the tide is **low**. As the earth turns, it carries the water with it, but not the bulges; these remain fixed along the earth–moon line. Under these circumstances, there would be a high tide every twelve hours and a low tide halfway in between. However, the moon is moving forward in its orbit, and it carries the tidal bulges along with it. The time interval from the moment when point a_1 is on the earth–moon line to M_1 until that point has made a little more than one revolution to point a_2 on the earth–moon line to M_2 averages 24^h50^m. At a_1 and a_2 the tide will be high, and there will be another high tide on the opposite side of the earth. Therefore, there will (on the average) be a high tide every 12^h25^m, with a low tide halfway between each high tide.

Solar tide may reinforce or partly cancel the lunar tidal range. The tide produced by the sun has an interval of 12 hours between successive solar high tides, but because the latter are somewhat less than half as high as the lunar high tides, the lunar tide predominates. The main effect is that of a high tide of variable height but with an interval of 12^h25^m between high tides. At new moon and full moon, when all three bodies are in the same line, the two sets of tidal bulges reinforce each other, and the range of water height is the largest. Such tides are called **spring** tides. At first quarter and last quarter phases of the moon, when the angle moon–earth–sun is 90 degrees, there is a partial cancellation of the lunar tidal range from high to low water and the range in height is the least. This is the time of **neap** tides. If at the time of spring tides the moon in its orbit around the earth is at perigee, the tide-raising force on the earth's water will be unusually high and the tidal range will be extremely large. When the moon is at apogee and the tide-raising force is the least, this combination with that of a neap tide will produce an unusually low tidal range.

The observed tidal phenomena are much more complex. In the open ocean, the spring tidal range is about 1 m (3 ft). Along a coastline, the nature of the tides is greatly affected by the slope of the ocean bottom, the shape of the shore line, the latitude of the place, and the time of year. In the Bay of Fundy between New Brunswick, Canada, on the north and Nova Scotia on the south, and in the deep inlets on the Patagonian coast inland from the southern Atlantic coast of Argentina, the funneling of water by the bays may cause a tidal range of 15 to 18 m (50 to 60 feet). It is interesting to note that the pull of the sun and moon produces in the solid part of the earth a tide with a range of about 23 cm (9 in). From this observation, it can be computed that the earth as a whole has a rigidity about twice that of steel.

2.10   THE PHASES OF THE MOON

The moon's phases are the result of the motion of the moon around the earth and the fact that moonlight is reflected sunlight. At **new moon**, in Figure 2.4, when the moon is nearly between the earth and sun, the illuminated half of the moon is toward the sun and we see no moon. **Full moon** takes place half an orbital period later (a little more than two weeks), and the situation is just the reverse of the one at new moon. At **first quarter** and **last quarter**, halfway between the new and full moon phases, we see only half of the illuminated surface, or one-quarter of the whole moon's surface. For a few days just after and just before new moon, when the moon is in the crescent phase, one can see that the dark portion is faintly illuminated. This

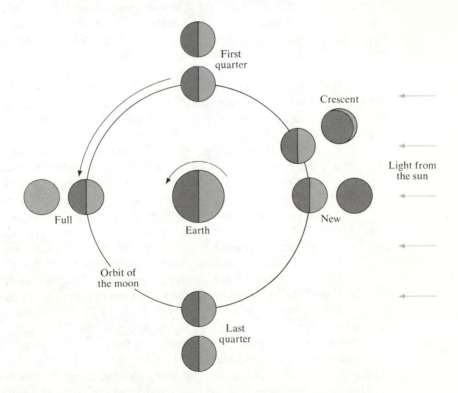

FIGURE 2.4 The phases of the moon. The circles on the moon's orbit are seen from well above the earth looking down on the earth and moon. The outer circles represent views of the moon as seen from the earth.

illumination is called **earthshine**, and it is caused by sunlight reflected from the earth to the moon and back again to the earth. The full moon is about nine times brighter than the quarter moon, even though only twice the illuminated area is seen at the full phase. This is because at first quarter we see many shadows cast by irregularities on the moon, but at full moon the object will hide its own shadow as seen from the earth. From new moon through first quarter to full the moon is said to be **waxing**, but from full through last quarter back to new moon it is said to be **waning**.

2.11 THE SURFACE OF THE MOON

As seen from the earth with large telescopes under the best of observing conditions, the moon's surface reveals an enormous amount of detail with great variety. Several tens of thousands of features have been mapped. Because the orbital period about the earth and the rotation period of the moon are the same, we always see the same side of the moon. The lunar exploration programs of the United States and the USSR have now photographed the whole lunar surface in detail.

The moon's surface is an interesting complex of mountains, craters, rays, ridges, narrow valleys, and plains. Some of the mountains reach altitudes of 7900 m (26,000 ft). About 30,000 craters have been mapped. Photographs from telescopes on earth show that the crater diameters range from 240 km (150 mi) down to the resolution limit of about 500 m (1500 ft). Closeup photographs taken from lunar orbiting vehicles and by the astronauts themselves on the moon reveal numerous craters down to 2.5 mm (1/16 in) diameter pits in surface rocks. Almost all the craters are nearly circular in outline, and some of the larger ones have rims up to about 3000 m (10,000 ft) high. See Figure 2.5. There are numerous rays radiating outward from a few of the largest craters. These rays are thought to be splashes of rock debris produced in the impact event that formed the crater. One of these rays appears to extend completely around the moon and back to the crater from which the material came.

The lunar plains, or seas, are called by the Latin word **maria** (plural, mah'-ree-uh; **mare**, singular, mah'-ray) because they were once thought to be bodies of water. However, it has been known for some time that the moon's surface has no liquid water and a negligible atmosphere without clouds or winds. The maria are relatively flat, smooth areas much darker than the rest of the moon. Some are hundreds of miles across and may easily be seen by the unaided eye, particularly at full moon. Most maria are circular in outline, but a few are irregular. Maria are distinguished by the low density of craters and mountains as compared with the rest of the moon.

Because the moon rotates once per revolution about the earth, a given surface point is in sunlight for about two weeks and then in darkness for the same time. The temperature of the center at full moon is about 133°C (270°F) but at lunar midnight two weeks later, the temperature drops to about − 156°C (− 250°F). [See Appendix 2 for the relationship between Fahrenheit, Centigrade (Celsius), and Kelvin (absolute) scales of temperature.] As the center of the full moon enters the earth's shadow during a total lunar eclipse, the temperature drops about 250 Fahrenheit degrees in one hour. This indicates that the heat conductivity of the surface material is very low and that it has a porous, sandy character. This has been confirmed by lunar surface missions.

FIGURE 2.5 A view of our moon taken by the spacecraft *Apollo 16*. In the upper right are three of the seas on the lunar near side. Most of the photo is of the lunar far side showing immense numbers of craters of many sizes. Solar illumination is from the upper right. (Photograph courtesy of the National Aeronautics and Space Administration.)

The figure of the moon is very nearly spherical with no clear-cut tidal bulge pointing toward the earth. The center of mass is within less than a mile of the moon's center.

2.12 THE ORIGIN OF THE MOON'S SURFACE FEATURES

It is important to understand that the moon has had no appreciable atmosphere for perhaps all of the billions of years since its origin. (See Section 4.13 for the escape of atmospheres.) Therefore, the erosional processes that have always been active on the earth have never operated on the moon, and the moon we see now is really a fossil body.

It is now generally accepted that all of the large- and medium-size craters (see Figure 2.6) down to about 30 m (100 ft) in diameter were formed by the impact of meteoroid bodies early in the history of the moon, when these bodies were far more abundant in the solar system than they are now. See Figure 2.6. Whatever their origin, these meteoroids were quickly swept up by the planets until now only a minute fraction of the original number remains. In the last 200 years no new craters

FIGURE 2.6 An *Apollo 16* photograph of the moon. Observe among this great complex of craters the flat-bottomed one (named Kohlschütter) above the center. This crater is characteristic of many on the far side. An instrument from the spacecraft protrudes from the right. (Photograph from the National Aeronautics and Space Administration.)

have been found, although some too small to be seen from the earth may have been formed. Similar craters must have been formed on the earth at the same time, but erosion and crustal movements have removed all traces of them. (More recent terrestrial meteor craters will be discussed in Chapter 4.)

Several characteristics of the lunar craters favor the impact theory. Outside and around many of the larger craters, the surface is very rough and appears to consist of the rock fragments ejected by the impact explosion. In most of the craters the inner floor is lower than a sizeable surrounding area outside the crater walls. The craters also have a large ratio of diameter to depth, which is not true for the earth's volcanic craters. These features appear to be results of an explosion in which relatively small bodies impacted at very high speed and penetrated to some depth. The enormous kinetic energy (energy of motion) would be dissipated in blasting out quantities of rock in all directions. Aerial bomb craters on the earth are almost always circular and are much larger than the bomb itself. Their shape does not depend on the impact angle of the bomb's trajectory, but rather on the explosion that takes place below the surface. This does not mean that there are no volcanic craters on the moon, but rather that this method of formation plays only a minor role. The mountains in many of the larger craters may be extinct volcanoes formed by the extrusion of molten rock from the interior through cracks in the crater floor.

The maria look like large lava flows, such as the extensive basalt rock flows in the Pacific Northwest. The relative absence of craters in the maria suggests that they were formed later in the evolution of the moon, after most of the meteoroids had been swept out of interplanetary space. In the maria are many deep valleys called **rilles**, which are about a half-mile wide, a hundred or so miles in length, and either straight or sinuous. The maria show extensive faulting, resulting in formation of long ridges and fault escarpments.

On rather rare occasions, minor disturbances are seen, such as the veiling or obscuration of a crater interior; most likely they are sudden outbursts of gas and dust from a newly opened crack in a crater floor. In 1958 a Soviet astronomer observed such an event and obtained a spectrogram of the light, which showed the presence of gaseous carbon. Most likely, the gas glowed because of the absorption of solar ultraviolet radiation and not because the gas was hot. These disturbances, often red in color, occur most frequently when the moon is at perigee (closest to the earth). The greater tidal effect of the earth would tend to open deep reaching surface cracks that are normally closed and release gas from deposits below the surface.

2.13 RECENT RESULTS OF LUNAR EXPLORATION

Since the 1960s the United States and the USSR have sent numerous space vehicles to explore the moon, but the United States has published more of its results. The most spectacular results were from the six Apollo missions in which manned spacecraft soft-landed on the moon. See Figure 2.7. The astronauts took many photographs, made many observations about the lunar surface, collected rock samples, and set up instruments which continued to send information back to the earth after the astro-

FIGURE 2.7 A photograph of the lunar surface taken by a crew member of *Apollo 15*. The lunar rover is parked near the edge of Hadley Rille. Note the numerous footprints made by the astronauts in the soft surface. (Photograph courtesy of National Aeronautics and Space Administration.)

nauts had left. The amount of information obtained by these means is staggering. In this section we will briefly summarize some of the more important findings.

Wherever the astronauts walked they left footprints about 0.6 cm (¼ in) deep in a layer of fine dust with the consistency of damp beach sand. The surface is strewn with rocks from pebble size to boulders 4.5–6 m (15–20 ft) across. Core drillings to a depth of a few feet show that the upper layers consist of fine dust and rock fragments of all sizes. It now appears that the outermost layer of the moon, called the **regolith**, consists of the dust, rock fragments, and boulders sprayed over the surface when the larger craters were formed.

In general, three types of rock material were found. The fine dust appears to have been formed in the impact crater explosions and thrown to great distances. (There is no air resistance on the moon.) Some of the dust consists of small glass beads which probably formed from the cooling in flight of sprays of molten rock from the impact craters. Some rocks showed mineral crystals as large as 6 mm (¼ in) across. Crystals of this size can only be formed when molten rock cools slowly. Therefore, these rocks were probably formed at considerable depth and were brought to the surface by large meteoroid impacts. Another rock type showed many *vesicles* (small pockets) in the rock of a kind seen on the earth in volcanic lava flows with a high dissolved-gas content. As the liquid rock reaches the surface, the pressure is released and gas bubbles form, which are quickly trapped when the rock freezes. One rock type, called *breccia*, is formed when rock fragments of more than one kind are cemented together by the heat and pressure generated by a meteoroid impact. All of the rocks were of igneous origin, which means that they had cooled from the molten state. No sedimentary rocks like sandstone or shale, which are formed on the earth by the deposition of water-borne sediments, were found.

Unlike those on the earth, none of the lunar rock samples contained any dissolved water. Core drillings to a depth of a few feet showed no permafrost. It was thought at one time that water erosion may have formed some of the moon's narrow meandering valleys, but it now seems clear that water has never been an important lunar erosional agent.

As compared with the terrestrial rocks, it is conspicuous that the lunar rocks are deficient in the lighter and more volatile chemical elements and overabundant in those that form minerals with high melting points. This suggests that much earlier the moon was very hot and that impact events often vaporized some of the local material. Because of the moon's low escape velocity (2.4 km/s or 1.5 mi/s), some of the lighter atoms might have easily escaped the moon's gravity while the heavier atoms did not. Numerous surface rocks show areas a few inches across covered by a thin layer of glassy material. These are probably splashes of molten rock from some nearby impact crater. Small glass-lined pits found on rock surfaces were most probably formed by the very energetic impact of small meteoroids. This may be the only erosional process now operating on the moon. Some of the oldest cratered areas on the moon appear to be somewhat smoothed—possibly the result of long exposure to bombardment by numerous small meteoroids.

The oldest rocks on the moon have a radioactive age of about 4.6 billion years. The maria are immense basalt flows made up of rock somewhat denser than that of the original surface. Radioactive dating shows that these flows were formed from about 3.9 to about 3.2 billion years ago. The liquid rock came to the surface through fissures a few hundred miles deep. It is thought that the rock was remelted by the heat generated by radioactive processes. The fissures were formed either by the normal cooling of the originally hot outer layers or by the impact of a large meteoroid. Standing on the edge of a rille in one of the maria, the astronauts could see layers of basalt flows intermingled with what appeared to be layers of debris ejected during the formation of craters.

Four seismographs were set up on the moon, the first by the *Apollo 12* astronauts. These instruments were capable of radioing to the earth the occurrence of a moonquake. When the *Apollo 12* astronauts left the moon and transferred to the orbiting command module, the lunar landing module was discarded, crashing back to the moon. The impact created a moonquake whose vibrations were detected by the seismograph for almost an hour, which is many times longer than the duration of the usual earthquake. About 3000 moonquakes are registered each year, but their intensity is feeble by terrestrial standards. Most moonquakes occur at a depth of about 700 km (450 mi). Most commonly these are caused by deep fault movements, but some may result from the impact of small meteoroids. The moonquakes reveal that the moon has a hot liquid, or partially liquid, core with a radius of about 500 km (300 mi).

It has also been found that the general magnetic field of the moon is much weaker than the earth's, and it differs a good deal from place to place because of the differences in the local magnetism of the rocks.

Much of our recent information about the lunar surface has been obtained from lunar orbiter satellites, unmanned space vehicles that have been placed into circular low-altitude orbits around the moon. They are able to take photographs of the lunar surface with high resolution and transmit these photos to the earth. It has been observed that when an orbiter is passing over any one of five of the seven circular maria, the orbiter first speeds up and then slows down as it leaves the area. This seems to indicate above-average gravity caused by excess density of the material in the mare. These mass concentrations are called *mascons* and, as said earlier, probably consist mainly of denser rock from the interior. However, this is still not perfectly clear. Another puzzle is that the far side of the moon does not have maria, but rather a number of large flat-bottomed *basins* ringed by one or more circular walls. The reason for the quite different appearance of the two sides is unknown. More recent work shows that there are no mascons on the far side.

2.14 THE HISTORY OF THE MOON

The debate over the moon's origin as an earth satellite is concerned mainly with three hypotheses: (a) the moon split off from a rapidly rotating earth, (b) the moon was captured by the earth after each had formed separately and elsewhere, and (c) the moon was formed together with the earth and in its present vicinity by the accretion of small bodies. The complexities of the arguments are extreme, so we shall be concerned only with a brief description of the probable process of formation of the moon alone.

It is quite possible that the moon was formed by the rapid accretion of meteoroids of all sizes slightly more than 4.6 billion years ago when, presumably, these bodies were far more numerous than they are now. The origin of these meteoroids is not clear, but some may have come from the asteroid belt (to be discussed in Chapter 4). As the meteoroids collided to eventually form the moon, they did so rapidly, and their kinetic energy heated the growing mass until it was hot enough to be molten or at least of a plastic consistency. In the fairly short time of perhaps 25 million years, most of the meteoroids have been swept up to form the inner four rocky planets and the moon, and the moon had nearly reached its present mass and had begun to cool and form a thickening crust. As soon as the hardened crust was sufficiently thick, additional meteoroid impacts produced permanent craters, but the impact rate was falling off very rapidly with time. Most likely, only a tiny fraction of the present craters were formed after 4.5 billion years ago. Even though gases like water vapor and those of the more volatile elements may have been released when the moon was hot, the low surface gravity could not retain them to form a permanent atmosphere.

The next step after the crater formation period was the formation of the maria, which began about 3.9 billion years ago and ended about 3.2 billion years ago. This process was discussed in Section 2.12.

It seems fairly obvious that the moon's surface experienced so many impact explosions that many parts of the crust have been scattered back and forth over wide areas. The present lunar surface seems to be covered by dust and rock fragments that came from somewhere else on the moon. It now seems safe to say that the lunar surface has remained almost completely unchanged for about the last three billion years.

2.15 ECLIPSES OF THE SUN AND THE MOON

Eclipses are basically very simple phenomena fraught with a wide variety of complications. Simply, a solar eclipse takes place when the sun, moon, and earth are in the same line (and in that order). A lunar eclipse occurs when the bodies are in the same line, but with the earth in between. Figure 2.8 diagrams some possible situations (not to scale). The shadows of the moon and the earth are right circular cones, and the apex of each cone is to the right of the body causing the shadow.

In the figure, the apex of the moon's shadow extends below the surface of the earth, and the intersection of the earth's surface with the dark cone is the **umbra**, the region of totality. Under the most favorable conditions, the maximum diameter of the region of **totality** is 260 km (167 mi). Because the earth's surface is curved and the moon's distance is variable, it is quite possible the shadow apex will not touch the earth and the eclipse will not be total. An observer on the shadow's axis (but beyond

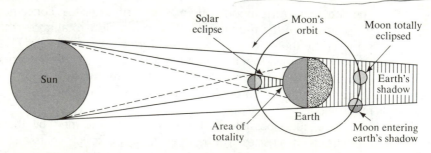

FIGURE 2.8 Eclipses of the sun and the moon (not to scale).

the apex) will see an **annular** eclipse in which the dark circle of the moon is surrounded by a narrow ring of sun. An observer in the region of partial shadow will see a part of the sun, and the eclipse will be **partial**. The area on the earth in which the eclipse is partial may easily be several thousands of miles across. Therefore, one's chances of seeing a partial eclipse are far greater than of observing a total one.

For a suitably placed observer, the duration of a partial eclipse may be a matter of hours, while the maximum duration of a total eclipse is 7^m39^s. The duration of totality depends mainly on the speed of the moon in its orbit and the rotational speed of the observer on the earth. The moon moves eastward in its orbit at an average speed of 3380 km/h (2100 mi/h). The maximum eastward rotational speed of the earth is 1700 km/h (1060 mi/h) at the equator. Therefore, the shadow moves eastward at the equator with a speed equal to the difference, or nearly 1600 km/h (1000 mi/h). At other latitudes, the earth's rotational speed is less, the shadow moves more rapidly, and the duration of the total phase is shorter. Sometimes it happens that the dark shadow cone misses the earth altogether, so no total eclipse is seen at all, but only a partial one.

A total eclipse of the moon occurs when the moon moves completely into the earth's dark shadow, which has an average length of 1,385,000 km (860,000 mi). When the moon moves along a diameter of the earth's shadow, the duration of the total phase is about 1^h40^m. It is preceded and followed by a partial phase about one hour in length. Sometimes the moon moves so that only a portion of its surface enters the shadow cone of the earth, and then only a partial eclipse is seen. Because of the refraction of sunlight by the earth's atmosphere and absorption in that atmosphere, some light is bent into the shadow, and the moon is illuminated by a faint, reddish glow.

2.16 THE ECLIPSE YEAR

From the last section it appears that one ought to observe a solar eclipse at each new moon and a lunar eclipse at each full moon. Obviously this does not take place, but why not? The main reason is that the moon's orbit is inclined by 5° to the plane of the earth's orbit (the ecliptic plane). Refer to Figure 2.9. In this figure the large circle is the earth's orbit about the sun. The smaller circle about each position of the earth (E) is the moon's orbit, but not to scale. The ecliptic plane is the plane of the page and the view is from the north side. Arrows on each orbit show the direction of motion of the body. For each circle for the moon's orbit, the solid part represents that part of the orbit above (north of) the ecliptic plane, and the dashed portion is that part of the moon's orbit below (south of) the ecliptic plane. In one revolution about the earth the moon will pass through the ecliptic plane twice at points called the **nodes**. In each case the letters *an* show the *ascending node* where the moon passes from the south to the north side of the ecliptic plane, and the letters *dn* refer to the *descending node* where the moon passes from north to south. The line joining the nodes and the earth's center is called the *line of nodes*. Figure 2.9 shows that in order to have either an eclipse of the sun or of the moon, the line of nodes must point toward the sun as it does at E_1, and the moon must be at one of the nodes.

For the time being let us assume that as the earth moves in its orbit (ccw) and carries the moon's orbit with it, the nodal line always remains parallel to itself. Let the

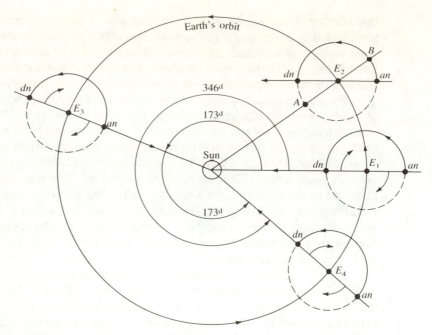

FIGURE 2.9 Diagram to explain the eclipse year.

next new moon occur when the earth is at E_2 and the moon at A. The moon will be south of the ecliptic plane, the moon's shadow will come to a point below the earth, and there will be no eclipse. If a full moon was to occur at point B, the moon would miss the earth's shadow because the moon would be well above the ecliptic plane. The next time when an eclipse could occur would be a half year after E_1 (not shown) when, once again, the nodal line would be pointing at the sun. This would mean that eclipses could occur only at six-month intervals, or twice a year.

Actually, over a period of some years, eclipses of either kind can occur on any day of the year. The nodal line does not remain parallel to itself, but revolves cw, with one complete revolution in 18.6 years. This means that when the nodal line points to the sun as it does at E_1, it will do so again 173 days later at E_3 and 173 days later at E_4. The **eclipse year** is 346 days, or the interval in time between E_1 and E_4 when the same node (in this case dn) is between the earth and the sun. It must be understood that even if the nodal line points toward the sun, it is necessary for the moon to be at the node for an eclipse of some kind. It should now be clear that averaged over a period of 18.6 years, eclipses of one kind or the other can occur on any day of the year.

The conditions for an eclipse as given above are a little too stringent, because when the moon is new the shadow of the moon is not directed toward a point in space but toward the earth, 8000 miles in diameter. At the time of full moon, the moon is trying to intercept a shadow nearly 6000 miles in diameter. For this reason the conditions for an eclipse of some kind may be relaxed somewhat to read that "for an eclipse of the sun or moon, the nodal line must point *close* toward the sun, and the moon must be *close* to one of the nodes." This provides for *eclipse limits*. It often happens that an eclipse of one kind is followed or preceded (or both) by an eclipse of the other kind. Note that in some years the nodal line may point toward the sun three times in one calendar year.

In any given calendar year, the least possible number of eclipses is *two*, both of the sun. The maximum number is seven, of which five will be solar and two lunar, or four solar and three lunar. The average number of eclipses of both kinds is about four.

2.17 THE IMPORTANCE OF ECLIPSES

Total lunar eclipses are now of little specific interest, except that temperature changes on the moon's surface can be measured at these times.

At one time, a total solar eclipse was of the greatest interest because only then could the prominences and the corona be seen. Because it is now possible to produce artificial eclipses at any time, the natural ones are less important. However, astronomers will go to scattered places on the earth to view (weather permitting) total solar eclipses in order to measure the bending of starlight by the gravitational field of the sun, to record the changes in the sun's radio noise during the eclipse, and to photograph the chromosphere at the beginning and end of each eclipse. These aspects of the sun will be discussed in a later chapter.

KEY TERMS

Foucault pendulum
ecliptic
zodiac
north (south) terrestrial pole
terrestrial equator
longitude
latitude
prime meridian
celestial sphere
zenith
altitude
azimuth
celestial equator
equinox

declination
right ascension
arctic circle

Tropic of Cancer
Tropic of Capricorn
precession
tidal friction
mean solar time
universal time
sidereal time
synodic month
sidereal month
chemical differentiation
earthquake waves

mantle
perigee
apogee

angular diameter
spring tides
neap tides
moon, phases
earthshine
lunar features
regolith
eclipses, sun and moon
lunar nodes
eclipse year

QUESTIONS

1. Strictly speaking, what is referred to as the earth's orbit around the sun is really the orbit of the center of mass of the earth-moon system. How does the observed eastward motion of the sun among the stars compare with what it would be if observed from the center of mass of the earth-moon system?

2. What is the terrestrial longitude of the North Pole of the earth?

3. Describe the nature of the earth's seasons under the following two hypothetical circumstances. First, the rotational axis is perpendicular to the plane of the earth's orbit; second, the axis lies in the plane of the orbit.

4. How would you expect the tidal range in the open ocean near the North Pole of the earth to compare with that in the equatorial regions?

5. How does the eccentricity of the moon's orbit alter the duration of a total eclipse of the sun in the case where the path of totality passes along the earth's equator?

3 The Tools and Methods of the Astronomer

Like any other scientist, the astronomer uses a variety of tools and techniques to obtain information and interpret data. Because almost all of this information is derived from the light from celestial bodies, it has been necessary to devise a wide range of instruments for the analysis of the light in all its colors and wavelengths. The three most important instruments used by the astronomer are the telescope, which gathers the light; the spectroscope, which analyzes the light into its various colors and wavelengths; and the photometer, which measures the intensity of the light.

3.1 LIGHT

Light is commonly thought of as the physiological sensation which results in vision and the phenomenon of color. In a much broader sense, light may be regarded as all forms of radiant energy which travel with the speed of light in a vacuum—about 300,000 km/s (186,000 mi/s). This broad definition covers such kinds of radiant energy as radio waves, infrared rays, visible light waves, ultraviolet light, X-rays, and gamma rays. All these forms of energy have the same speed in a vacuum. A more technical term for these several forms is **electromagnetic radiation**. The main difference between these various forms is wavelength. The accepted unit of wavelength for much of this radiation is the **angstrom (Å)**, which is 1/100,000,000 of a centimeter. One centimeter is nearly 0.4 inch. The approximate range of wavelengths for visible light is from about 4000 Å for the lower limit of violet light to about 7500 Å for the longest red rays. The normal human eye has its greatest sensitivity in the yellow-green region at about 5500 Å, or about 2/100,000 of an inch. Wavelengths shorter than 4000 Å are in the ultraviolet (UV) region. At about 10 Å, the X-ray region begins. At still shorter wavelengths of less than about 1 Å, the gamma rays are found. Waves longer than 7500 Å are in the infrared region, a very broad one, up to about one centimeter (100,000,000 Å). At about this limit the radio region begins; here the wavelengths are given in centimeters or meters. One meter equals 100 centimeters, or about 39.4 inches. None of the regions is sharply bounded.

3.2   REFLECTION, REFRACTION, AND DISPERSION

In Figure 3.1, consider a narrow beam of light consisting of the colors violet (V) and red (R). Let this beam in air strike a flat glass plate at some angle other than zero with respect to the perpendicular (the **normal**) to the surface. As seen in the figure, three things will take place. First, a part of the incident beam will be reflected from the air-glass interface. The law of reflection states that the angle of incidence (i) equals the angle of reflection (i') and that the normal is contained in the plane of the incident and reflected rays. This is true for all colors or wavelengths.

The second effect is that after entering the glass, the direction of the beam will be changed in such a way that it is bent *toward* the normal. This is called **refraction**. The third effect is **dispersion**, in which violet light is bent more than red. The angle of refraction r (measured from the normal) is less for the violet light. Intermediate values of r will be found for the colors between violet and red. The angle of refraction depends upon the angle of incidence (i) and the **index of refraction** (n). The latter is the ratio of the speed of light in air (close to the speed in a vacuum) to that speed in some transparent medium such as glass or water. The index (n) is always greater than unity. The greater the value of n, the greater is the change in direction of the beam; that is, the smaller the angle r. For all transparent media, n will increase as the wavelength decreases, and the variation of n with wavelength will be different for each medium.

3.3   THE PRISM AND THE SPECTRUM

Consider a beam of white light (all colors) which strikes a prism with a prism angle A, as in Figure 3.2. As the light enters the glass, it will be split into separate rays for each component color or wavelength (only two are shown in the figure). A further bending will occur when the rays leave the prism. If a white card is placed in the emergent dispersed beam, a spectrum will be seen. A **spectrum** is a display, or analysis, of the component colors, or wavelengths, in the original beam. Very commonly, the spectrum is photographed for detailed study, but other instruments, such as various types of photoelectric cells, are also used. The spectrum of a source commonly extends into the ultraviolet and infrared regions, and these are studied in the same way.

FIGURE 3.1   Reflection, refraction, and dispersion.

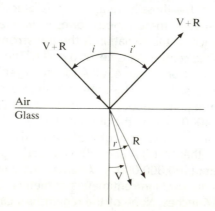

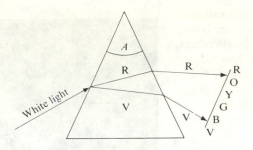

FIGURE 3.2 The prism and the spectrum.

3.4 LENSES

With the above background we are now in a position to understand the action of lenses. Almost always, the surface of a lens is either a plane surface or a section of a sphere. Figure 3.3 shows the action of a double convex lens on parallel light rays that are parallel to the lens axis and are of two colors, say, violet and red. After the parallel rays from the left pass into the lens and emerge on the right, they are refracted and converge to a point on the axis called the **focus**. There is a different focus for each color. One can regard a lens as a prism with a continuously variable prism angle. On the axis, the prism angle is zero, and it increases continuously to its maximum value at the edge of the lens. Because the index of refraction for violet light is the greater, the distance from the lens to its focus (the **focal length**) is less than for red light.

Before considering image formation by lenses, it would be well to comment on the fact that the lens in Figure 3.3 does not have the same focus for violet as for red light. The other colors have intermediate foci. This effect is called **chromatic aberration**, and it is one of the several faults, or aberrations, of lenses. The ideal lens would bring all the rays to the same focus. Other aberrations not to be discussed here have names such as spherical aberration, astigmatism, coma, etc. For their study, the student is referred to a text on geometrical optics. It is not possible to eliminate all aberrations completely, but by proper lens design and the use of more than one lens in the system, one can reduce all of them to minimum values. Chromatic aberration can be greatly reduced by a combination of two lenses of different kinds of glass.

When a lens is used as the objective (larger) lens of a telescope, it is mounted at

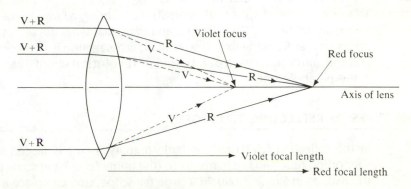

FIGURE 3.3 The action of a lens on parallel light rays of two colors.

FIGURE 3.4 The 36-inch refractor of the Lick Observatory is located on 4200-foot Mt. Hamilton, some 20 miles east of San Jose, California. It is operated by the University of California, Santa Cruz. (Lick Observatory photograph.)

one end of a tube whose length is the focal length of the objective lens. At the other tube end the smaller eyepiece lens is mounted so that its focus coincides with that of the objective lens. The magnification of the telescope will be the focal length of the objective lens divided by the focal length of the eyepiece. See Figure 3.4.

3.5 THE REFLECTING TELESCOPE

One good way to completely eliminate chromatic aberration is to use a mirror telescope, as described in Figure 3.5. Let parallel rays from the left strike a mirror whose surface is a section of a sphere, and let the center of curvature be the point c on the mirror axis. Observe that the innermost and outermost sets of rays are not reflected to the same focus. For a set of rays very close to the axis, the focal point will be halfway between the center of curvature (c) and the mirror surface. For parallel rays farther from the axis, the focus will be at points progressively closer to the mirror. This effect is called **spherical aberration**. To bring all rays to the point c/2, it is only necessary to increase, in a smooth and particular way, the radius of curvature of the zones farther from the axis. The resulting cross section is not a circle but a **parabola**, and the mirror surface is a **paraboloid of revolution** about its axis. The dashed curve is the parabolic one.

3.6 THE TYPES OF REFLECTING TELESCOPES

In the reflecting telescope, the large primary mirror is placed at the bottom of a supporting tube and the mirror focus (the prime focus) is at some point in the tube on the axis, as in Figure 3.6(a). In a large telescope, one can place a plate holder at the

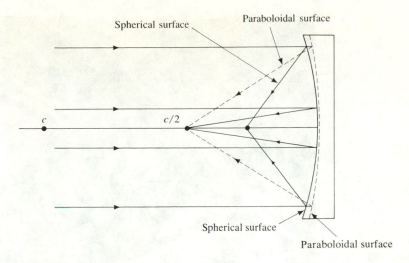

FIGURE 3.5 Reflection from a spherical and from a paraboloidal surface.

prime focus; for a few large reflectors, such as the 200-inch Hale telescope (shown in Figure 3.7), a cage for the observer is placed at the upper end of the tube near the prime focus. The cage obstructs some of the light but does not spoil the image formation. For smaller telescopes, where one might wish to use an eyepiece, a flat mirror is placed in the tube before the focus so that the focal plane is brought outside the tube. This is the Newtonian form, as shown in Figure 3.6(b). The third standard form is the Cassegrainian [Figure 3.6(c)]. Before the light reaches the prime focus, it is intercepted by a hyperbolic mirror which reflects the light back down the tube through a hole in the mirror to a point behind the mirror. This form is convenient to use, because the focus is always closer to the observing floor than it is in the Newtonian.

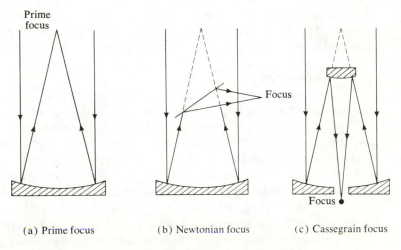

FIGURE 3.6 The primary types of arrangements for reflecting telescopes.

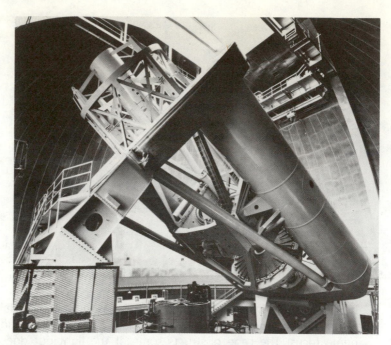

FIGURE 3.7 The 200-inch (508-cm) Hale reflecting telescope pointing north in its 186-foot-diameter dome on Mount Palomar east of San Diego, California. At the lower end of the open tube is the covered mirror; the cylindrical observing cage is at the other end of the tube at the upper left. (Palomar Observatory photograph.)

3.7   REFRACTING AND REFLECTING TELESCOPES CONTRASTED

The primary reason for using a telescope is to gather light and to increase the brightness of an object with respect to what it would be to the naked eye. Magnification is usually a secondary consideration. For many years, almost all observations have been made using photography. Visual inspection is sometimes necessary and even superior to the photographic record, as in observations of Mars. But the great advantage of a photographic record is that it is permanent, and the photographic plate contains far more information than can be gathered by the eye and remembered by the brain.

The larger the telescope objective, the more light is gathered. Fainter objects can be detected and brighter ones can be photographed in a shorter time. It is difficult to obtain large lenses free of defects, such as inhomogeneities in the glass and small bubbles, for a refracting telescope. For a two-element refractor, four surfaces must be polished and figured to the correct curve. For a reflector, there is only one surface to be figured and the glass need not be transparent. Most lenses absorb the ultraviolet part of the spectrum below about 3400 Å and also large parts of the infrared region. For a glass mirror coated with aluminum, the range of the reflected spectrum is far greater. Since the glass of a mirror need not be transparent, large thick mirrors can be made to minimize the bending. A lens can be supported only around its edges, but a mirror has a system of edge and back supports that may almost totally eliminate bending (flexure).

The first telescope mirrors of the late seventeenth century were made of *speculum* metal (an alloy of several metals), which could be given a high polish but which tarnished rather easily. To remove the tarnish the whole mirror had to be refigured—a tedious job. In the latter half of the nineteenth century mirrors began to be made of ordinary glass because, in part, it was discovered how to chemically deposit a bright silver coat on the curved glass surface. Silver also tarnishes rather easily, but it is not difficult to remove from the glass and resilver. In the 1930s silver was replaced by a vacuum-evaporated coat of aluminum, which is harder than silver, a better reflector in the ultraviolet, and more resistant to tarnish.

The main difficulty with ordinary glass mirrors is the image distortion that arises as the air and the outer parts of the mirror cool at night. Ordinary glass changes its shape considerably with rather small temperature changes. Pyrex is a much better material because its expansion rate with temperature is much less than that for glass, and quartz is even better. About 1960 the technical problems of casting this high melting point material were solved. Just a few years later new ceramic materials were developed which exhibit almost no change of shape with temperature. Now all large mirrors are made of the ceramic material, and telescope performance has improved greatly.

The largest reflector in the United States is the 200-inch Hale telescope on Mount Palomar in southern California. The largest reflector in the world is one with a diameter of 600 cm (236 in) made by the USSR and installed at a site in the Caucasus Mountains.

3.8   TELESCOPIC RESOLUTION AND ATMOSPHERIC TRANSPARENCY

There is a limit to the amount of detail any telescope can reveal, regardless of the amount of magnification used. This limitation is illustrated by the inability of a given telescope to detect lunar features below a certain size or to reveal the duplicity of a pair of stars at less than a given angular separation. The basic limitation is the finite wavelength of light, and the explanation is one that the student must find in a text on physical optics. **Resolution** is the minimum angular separation at which a given telescope can resolve a source into two light points rather than one. This angle is made smaller as the aperture of the telescope is increased and also as the wavelength becomes shorter. Wavelength is not important for visual purposes because the range is small, having an average value of about 5500 Å. For visual telescopes using the average wavelength of 5500 Å, the minimum resolution angle in arc-seconds is given by $4''.5/d$, where d is the aperture of the telescope in inches. Thus, if the telescope diameter is 102 cm (40 in), the minimum angle is about 0.1 arc-second. To repeat, this means that a 40-inch telescope can reveal that a given star is really two close stars only if the angular separation between the two stars is about $0''.1$ or larger. For the 200-inch Hale telescope, the minimum resolution angle would be $4''.5/200$, or about $0''.02$.

The above values are theoretical values. For the 200-inch telescope, the value $0''.02$ will seldom if ever be realized because the light must pass through the earth's atmosphere. When there are erratic temperature changes in the air, the image becomes enlarged and subject to rapid vibrations. Thus, on a night of poor ''seeing,'' the minimum resolution angle will be much larger than the theoretical value for a given telescope. Astronomers spend much time on site surveys locating places where meteorological conditions permit the best conditions for the maximum number of viewing hours.

Another difficulty with the atmosphere is that while it is almost completely transparent between about 2900 Å in the ultraviolet and about 12,000 Å in the near infrared, it is largely opaque outside this range. It is completely opaque below 2900 Å due to the absorption by ozone and oxygen molecules. Beyond about 12,000 Å in the near infrared, the atmosphere is largely opaque except for occasional "windows" where an appreciable amount of radiation can get through. Here the absorption is largely due to water vapor and carbon dioxide. Thus it appears that the best place to avoid this absorption is to observe from altitudes of at least 160 km (100 mi), where the amount of remaining atmosphere is negligible. Some observing is done from rockets, which in each flight spend a few minutes above this altitude, but most is done from spacecraft orbiting the earth. These spacecraft radio back the intensity of ultraviolet light, X-rays, and gamma rays from all over the sky. Some of these discoveries will be discussed in the coming chapters. With their several kinds of telescopes, these spacecraft are expensive to build, launch, and operate, and their lifetime in space is rarely more than one or two years. But there is no other way if we want to study those regions blocked by the earth's atmosphere. However, this does not mean that earth-based astronomical observatories are going out of style; they, too, have their place and will serve us for a much longer time.

3.9 SPECTROSCOPY

Spectroscopy is the science and art of the analysis of light by means of a study of the light's component wavelengths (the spectrum). Astronomical spectroscopy is a highly sophisticated field, and its study has revealed an enormous amount of information about our solar system and the stars beyond.

The spectrum is produced by a **spectroscope**, a simple form of which is shown in Figure 3.8. Light from the telescope enters a narrow slit on the left. The diverging rays from the slit are intercepted by a collimator lens and made parallel. Then they pass through the prism and are dispersed. The camera lens focuses the dispersed rays

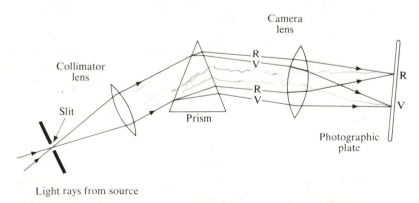

FIGURE 3.8 A simple spectroscope showing the formation of two separate spectral lines, one each for a wavelength in the red (R) and the violet (V).

on a photographic plate. This instrument is called a spectroscope when the spectrum is observed visually and a **spectograph** when the spectrum is photographed. If the slit is illuminated by light of only one wavelength, an image of the slit is produced at the appropriate place in the spectrum. This image is called a **spectral line** and is an image of the slit in monochromatic light.

3.10 THE THREE TYPES OF SPECTRA

The Continuous Spectrum

Fortunately, all kinds of spectra may be reduced to three classes, even though they may and often do occur in combination in astronomical cases. The first is the **continuous spectrum**, which consists of all wavelengths. This type is emitted by any hot solid or any hot or glowing gas under high pressure, regardless of its chemical composition. To the eye, the appearance of the visual region is a continuous band of color from violet through red, although the ultraviolet and infrared wavelengths are also present. Two sources of a continuous spectrum are the hot tungsten filament of a light bulb and the hot high-pressure gases of the sun's photosphere.

The only difference between the continuous spectra from various sources depends upon the source temperature. In Figure 3.9, the spectral energy distributions for the sun (5750K) and for two other sources, one much hotter and the other much cooler, are plotted. [See Appendix 2 for the Kelvin (K) scale of temperature.] The Kelvin scale will be used commonly in our later discussion of stellar temperatures. The

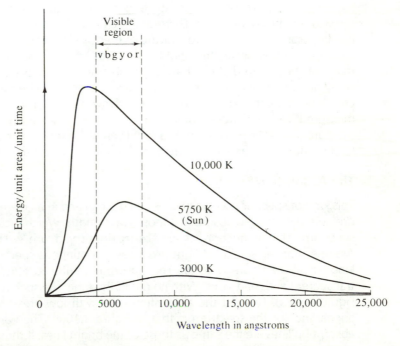

FIGURE 3.9 The spectral energy distribution curves for the sun and for two other bodies, one hotter and the other cooler.

theoretical curves for each temperature are called *black body radiation* curves. Along the vertical axis are expressed energy units per unit area per unit time. An example might be calories (or ergs) per square centimeter per second. The energy is zero at zero wavelength. For any temperature, as the wavelength increases from zero, the energy rises to a maximum and then decreases and approaches zero again as the wavelength approaches infinity. It can be shown that the total energy in all wavelengths (the total area under the curve) increases as the fourth power of the absolute temperature, and also that the wavelength of the maximum varies *inversely* as the first power of the absolute temperature.

Note the part of each curve which lies in the visible region between the two dashed vertical lines. For the cooler body, there is more red than blue-violet light; therefore, its color will be predominantly red. For the hotter body, there is more blue-violet than red light, and its color will be on the blue side. For the sun's curve, there are about equal amounts of radiation from either end of the visible region, so the resulting color is close to white or perhaps a little on the yellow side. We shall refer to these curves again.

The Bright Line Spectrum

This second type of spectrum is emitted by any hot or glowing gas under *low* pressure. The spectrum from any such source is characterized by a group of spectral lines, each of a very narrow range of wavelength or color. There is no light between the lines—the spaces are dark. Figure 3.10(b) shows the **bright line spectrum** in the visible region of mercury vapor under low pressure. The important feature of this kind of spectrum is that every chemical element and each of its possible stages of ionization has its own characteristic bright line spectrum. No two are alike. None of the wavelengths is the same. Therefore, the bright line spectrum of an atom becomes its "fingerprint" and serves to indicate its presence in the source. After a great deal of labor by spectroscopists, the bright line spectra of all of the chemical elements and of many of their ionized states have been obtained; the wavelengths have been measured and are available. In theory, only one line needs to be measured to identify the chemical element from which it comes. This would be true if wavelengths could be measured with infinite accuracy, but because they cannot, it is the practice to measure several lines for identification. Even if the source consists of more than one kind of atom and the spectra are superimposed, the identification can still be made.

The Absorption Spectrum

This kind of spectrum occurs because, in general and in an elementary way, each low-pressure gas tends to absorb just those wavelengths which it can emit as a bright line spectrum. The production of an **absorption spectrum** is shown in Figure 3.11. Assume that the source of light produces a continuous spectrum; that is, all wavelengths, as in Figure 3.10(a). The absorption tube just in front of the slit has transparent ends and sides. With no gas in the tube, the spectrum is continuous, as in Figure 3.10(a). Now fill the absorption tube with low-pressure mercury vapor. The appearance of the spectrum is that in Figure 3.10(c). The wavelengths of the dark absorption lines are the same as those of the bright lines. It thus appears that mercury vapor acts as a filter in the sense that it is opaque to the light of the wavelengths of

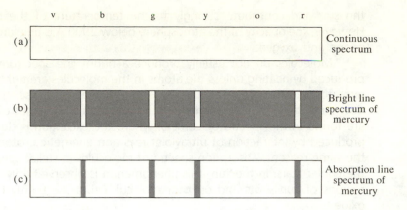

FIGURE 3.10 The three types of spectra.

the bright lines and transparent to all others. Therefore, the absorption spectrum of an element also identifies the element.

When the atoms in the tube absorb certain wavelengths, the energy removed from the beam is stored up in the atoms, but it is quickly released. The direction in which the light is released is random with respect to the direction from which the light came from the continuous source, so only a very little is reradiated toward the slit. Now, if one were to point another spectoscope toward the side of the absorption tube, the bright line spectrum of mercury would be seen. The mercury vapor in the tube absorbs just those wavelengths which it would radiate as a low-pressure gas. If all the reradiated energy continued on through the slit, there would be no absorption lines. The manner of formation of the absorption spectrum of mercury applies to any other gas at low pressures.

The production of an absorption spectrum in the laboratory is usually not as simple as indicated here, but in principle the description is correct. Some gases absorb at room temperatures and others only at elevated temperatures.

Molecules may also absorb, and each kind of molecule has its own characteristic absorption spectrum. Most molecular absorption is in the infrared region. Molecules in our atmosphere, such as oxygen, nitrogen, carbon dioxide, and water vapor (particularly the latter two), absorb fairly large quantities of the infrared radiation of

FIGURE 3.11 The production of an absorption spectrum.

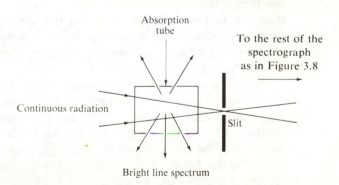

the sun and contribute strongly to the temperature of the earth's atmosphere. However, the opacity of the atmosphere below 2900 Å is due largely to absorption of ozone and oxygen.

Molecules do not usually produce a bright line spectrum when excitation is produced by heating unless the atoms in the molecules are tightly bound. When the interatomic forces are weak, the elevated temperatures cause the molecules to break down into molecular fragments or individual atoms. Examples of molecular emission are found in the spectra of comets, where the excitation is due not to heat but is produced by interaction of ultraviolet rays and energetic protons from the sun. For the same reasons, absorption spectra of molecules in stellar spectra are seen only in the cooler stars. In the sun, this phenomenon is observed only in the cooler central regions of sun spots and only for molecules that are tightly bound, like titanium oxide.

An understanding of these laws of spectra is very important to the astronomer. A determination of the shape of the spectral energy distribution curve for a star will lead to a determination of its temperature. The identification of the wavelengths of the absorption or emission lines will reveal which elements are present. The absorption spectrum of a planet's atmosphere will help define its chemical composition. The presence in a star's spectrum of many absorption lines due to ionized elements indicates that its temperature is high. We shall refer to these types of spectra often.

3.11 THE ELEMENTS OF THE ATOMIC THEORY

In order to make the three kinds of spectra more meaningful, it is desirable to discuss the way in which atoms produce absorption and emission spectra. The theory of atomic spectra can be very complex. A simplified explanation is presented here and in the next two sections.

Basically, an atom has two main parts—a **nucleus** with a positive electrical charge and a cloud of negatively charged **electrons** swarming around the nucleus. The central nucleus is about ten thousand times smaller than the whole atom and is made up of particles called **protons** and **neutrons**. The proton is a positively charged particle, and the neutron has no electrical charge. Their masses are nearly the same, and each has a mass about 1840 times greater than that of one electron. The quantity (or strength) of the electric charge on a proton and an electron is the same, but their signs are opposite. In most atoms there are more than one **isotope**, each of which has the same number of protons but a different number of neutrons and, of course, for the neutral atom the same number of electrons as protons. The chemical properties of all the isotopes of a particular atom are nearly, but not quite, identical.

All atoms of a particular chemical element have the same number of protons in the nucleus. Except in the simplest form of hydrogen, all nuclei have neutrons, but their number does not concern us here. There are no electrons in the nucleus. In a **neutral** atom, the total amount of positive electrical charge in the nucleus is exactly balanced by the total amount of negative charge on the electrons. Thus, in a neutral atom, the number of protons and electrons is equal. It is a relatively difficult matter to remove a proton from a nucleus, but it is rather easy to strip an electron from an atom. The process by which an electron is removed is called **ionization**. When one electron is removed, the atom is said to be singly ionized; when two are removed, it is doubly ionized; and so on for higher states of ionization. Atoms can be ionized by the collision of high-speed atoms in a hot gas, by bombarding the atoms with energetically moving electrons from outside, or by exposing the atoms to sufficiently energetic light radiation. The last method will be discussed later.

The simplest atoms are hydrogen atoms with one proton in the nucleus and one electron revolving about the nucleus. The next more complicated atom is helium, whose nucleus contains two protons (and some neutrons) around which revolve two electrons. The most complicated atom found in nature (not artificially produced in the laboratory) is uranium, whose nucleus contains 92 protons matched by a cloud of 92 electrons. In a simple way, one may regard an atom as a solar system in which the massive nucleus is the sun and the electrons correspond to the planets. Because an atom is a submicroscopic particle, it is not possible to actually *see* what it looks like, but one can make models based on the observed behavior of the atom.

3.12   A MORE EXTENDED VIEW OF THE NATURE OF LIGHT

Light may be regarded as a wave motion much like water waves. This idea is perfectly satisfactory in explaining a wide range of phenomena, like reflection and refraction (Section 3.2); but a variety of effects involving the interaction of light with atoms cannot be explained by the wave theory. About A.D. 1900, there arose a new theory—the quantum theory (quantum from the Latin for quantity)—which clarified these earlier observations. The theory states that a beam of light is not a succession of waves, but rather is a stream of particles of energy all moving with the speed of light. These particles are sometimes called quanta (plural of quantum), but a more commonly used term is **photons**. In the wave theory, one regards the light energy from a source as consisting of a continuous stream of waves, such as those produced when a rock is tossed into a quiet pool of water, and the water waves radiate from the impact point in concentric circles. In the quantum (photon) theory, the light energy proceeds in all directions from the source (such as the filament of a light bulb) as a random stream of individual particles of light energy. An imperfect but reasonably accurate comparison would be a lawn sprinkler, which sprays the grass with individual water drops.

The amount of energy contained in a photon depends in a simple way on the wavelength of the light. It may seem odd to speak of the wavelength of a photon, but it is convenient to do so in order to relate the quantum and wave theories of light. The relation is given by the formula $E = C/\lambda$, where E is the photon's energy, λ the wavelength of the light, and C is a constant whose value depends on the units used for energy and wavelength. In our discussion, we need not be concerned about the units. We need only to realize that since λ is in the denominator of the fraction, the longer the wavelength of an individual photon, the lower will be its energy. Conversely, the shorter the wavelength, the larger and more energetic will be the photon. Therefore, if we were to proceed to shorter wavelengths in going from infrared to red to violet and on to even shorter wavelengths in the ultraviolet and X-ray regions, the photon energy would increase.

Both wave and quantum theory are correct, and each has its appropriate use.

3.13 ENERGY LEVELS IN ATOMS—THE ABSORPTION AND EMISSION OF LIGHT BY ATOMS

We are now in a position to consider the mechanism by which atoms absorb and emit light. To do this, we shall consider a model of the simplest of all atoms—the hydrogen atom. As was said previously, this atom is made up of a nucleus with a single proton around which revolves a single electron in what may be regarded as a circular orbit.

Imagine it is possible for the electron to move in an orbit of any radius. Remember that because the electrical charges of the proton and the electron are of opposite sign, there is an attraction between them. As a start, let the electron be moving in a small orbit close to the nucleus. Then, by some means, it is moved into a larger orbit. Because of the attractive force between proton and electron, it is necessary to oppose this force when the electron is moved into a larger orbit. Work must be done on the electron, and energy must be given to the atom as a whole. As a result, the electron will have more energy of position in a large orbit than in a smaller one. A reasonable analogy is a person walking up a ramp. At the top of the ramp, the person is in a higher state of energy of position because work was done in increasing his distance from the center of the earth against gravitational attraction.

It had been discovered through experiment and mathematical studies that the electron cannot exist in an orbit of just *any* radius. In a simple way, the size of the orbits is restricted by the forms $r_n = r_1 \times n^2$, where r_1 is the radius of the smallest permitted orbit and r_n is the radius of the nth orbit, where $n = 1, 2, 3$, etc., up to any value of n, although there are practical limits. The electron energy is the lowest in the first orbit and increases by distinct steps as it is raised to higher orbits. One may say that there exists for the electron—or for the atom as a whole—a series of energy levels, where E_1 is the energy of the lowest orbit, E_2 of the second orbit, and so on. When the electron has the energy E_1, it is said to be in the **ground state**, and in higher levels, it is in **excited** states. The diagrams in Figure 3.12 are an illustration of the energy levels for hydrogen. The vertical scale is in order of increasing energy upward. Observe that as one goes upward, the energy difference between successive levels decreases. The convergence is such that the energy does not increase indefinitely but finally reaches a level $E\infty$ (read E infinity) beyond which there are no definite levels. The energy $E\infty$ corresponds to that of an electron orbit in which n approaches

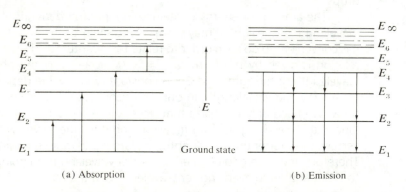

(a) Absorption   (b) Emission

FIGURE 3.12 The energy-level diagram for the hydrogen atom.

infinity. The orbit cannot be infinitely large, but n reaches some very large value. For practical purposes, the orbit radius cannot be larger than the distance between the atoms in a gas. For our purposes, this distance is great enough so that by the standards of the hydrogen atom it is essentially infinitely large. $E\infty$ is the energy required to completely remove the electron from the hydrogen atom; that is, to ionize it. All of the possible energy differences between states have different values—none are identical.

We are now in a position to understand the absorption and emission of light by the hydrogen atom. Let us start with the electron in the ground state E_1. If the electron is to be raised to the second orbit, or second state of energy E_2, one must supply (to the atom) the energy equal to the difference $E_2 - E_1$. One way to do this is to have the atom absorb a photon having just this amount of energy. If the photon energy were just a little bit less (or a little more) than $E_2 - E_1$, the atom would ignore the photon, and it would not be absorbed. If the photon energy is increased to $E_3, - E_1$, the atom would again absorb the photon, and the electron would jump from the first to the third orbit. Therefore, for hydrogen atoms in the ground state, the gas is transparent to all photons except those with the energies $E_2 - E_1$, $E_3 - E_1$, $E_4 - E_1$, etc. The electron has no place to go and can have no existence between the orbits corresponding to this series of energy levels.

When the electron is in one of the excited states, it will remain there for about one one-hundred-millionth of a second and then drop back to the ground state. If, while it is in an excited state, the atom receives a photon whose energy corresponds to the difference between that state and some higher state, the electron will jump to the higher energy state, or level. Figure 3.12(a) shows jumps from the ground state to E_2, E_3, and E_4, and also a transition from E_4 to E_6.

Previously, it was seen that the relation between energy and wavelength for a photon is $E = C/\lambda$. By inversion, the formula becomes $\lambda = C/E$. The wavelength of a photon that causes the transition from the ground state to E_2 would be $C/(E_2 - E_1)$, and so on, for any other energy difference. If light from a continuous source of radiation were to pass through hydrogen, the consequent spectrum would show absorption lines at just those wavelengths corresponding to photons whose energies were equal to all of the possible energy differences in the energy-level scheme of hydrogen.

Now consider the case of a hydrogen atom in some excited state in which the electron is not to be raised to some higher level. After its short lifetime in that level, it will leave and cascade back down to the ground state. As it cascades down, it becomes de-excited and can arrive at the ground state in different ways depending on the level from which it starts. Let the electron be in E_4, and see Figure 3.12(b). The electron could make one large jump all the way down to the ground state and emit a photon with energy $E_4 - E_1$. The other possibilities are E_4 to E_2, E_3 to E_2, and finally, E_2 to E_1 with the emission of three successive photons whose energies would be, respectively, $E_4 - E_3$, $E_3 - E_2$, and $E_2 - E_1$. Further possibilities are E_4 to E_3 and E_3 to E_1, or E_4 to E_2 and E_2 to E_1. If the atom started at some higher state than E_4, the possibilities become more numerous. Quite naturally, one atom can exhibit only one pattern at a time. If it were excited back to the same state again, it might well take a quite different route back down to the ground state. In the practical case, many hydrogen atoms are involved. Each one acts and behaves like an individual atom, with the result that all possible transitions could take place. When this emitted radiation is observed by a spectroscope, one sees an *emission* spectrum, and it will contain just those wavelengths that would be missing in the absorption spectrum.

This explanation is a simplified version of the phenomena of absorption and emission of radiation by hydrogen gas. For any other chemical element in the gaseous state, the explanation is much the same, but the energy-level diagram is more complex. Remember that the absorption and emission spectra of the neutral atoms of a particular chemical element and those for the various states of ionization are all different. This is because their energy-level arrangements are different. As a consequence, it is possible to tell from the appearance of the spectrum what the various stages of ionization in the gas may be.

Brief mention should be made here of the emission of a continuous spectrum by a hot gas under high pressure and a hot solid. In both cases, the atoms are so close together that they disturb each other to an extreme extent. This causes the normal energy levels to broaden to the point where they overlap, and the well-defined energy levels no longer exist. This result is that the normal set of definite levels is broadened out into a continuous band containing, in effect, an infinite number of levels. Therefore, transitions of all energy sizes can take place, and all wavelengths will be present to form a continuous spectrum.

3.14 THE DOPPLER EFFECT—RADIAL VELOCITY

When one listens to a tuning fork vibrating with a frequency of 250 vibrations per second, a note of that **pitch (frequency)** is heard by the ear. In effect, the ear "counts" the number of sound waves that reach it per second, and the sensation of pitch is produced. Sound waves are a succession of air pressure changes moving outward in spherical waves with the source as center. These waves cannot be seen, but a two-dimensional representation would be much like what is observed when a stone is dropped into a pool of quiet water. Here one sees circular crests and troughs moving outward from the point of the stone's impact. In air at normal temperatures, sound travels about 310 m/s (1000 ft/s). In a train of waves 1000 feet long, there would be, in this case, 250 waves. Hence, the wavelength of the sound waves would be 1.2 m (4 ft). Therefore, the wavelength (λ, lambda) is equal to the speed of sound (v) divided by the frequency (f); that is, $\lambda = v/f$ or $f = v/\lambda$, and $\lambda \times f = v$, where v is a constant value.

Consider a stationary sound source emitting sound waves at a given frequency, wavelength, and wave velocity and an observer who is moving *away* from the sound source at a speed less than that of sound. The result will be that the observer's ear will count fewer waves per second; that is, he will detect both a lower frequency than that emitted by the source and a longer wavelength. If the observer were moving toward the source, he would count more waves per second (a higher frequency) than the source frequency and the waves would seem to be shorter. The same would be true if the source were moving toward a stationary observer. For example, consider an observer listening to the siren of an approaching ambulance. The moment that the ambulance passed him, the pitch would drop below the pitch observed when the ambulance was approaching. It does not matter whether the source or the observer are moving or if both are moving. It is the relative motion that matters. If they are receding from one another, the observed frequency will be less and the wavelength longer; and, if they are approaching each other, the result will be just the reverse. This *apparent* change in frequency of the true value emitted by the source due to the relative motion of source and observer is called the **Doppler effect**. The effect is not to be confused with the change in intensity or loudness of the source resulting from distance changes.

In very much the same way, light changes in wavelength and frequency when we consider the relative motion of the source of light (a star) and an observer on the earth using a telescope and a spectrograph. When the star and the observer are receding from one another, all the wavelengths of the absorption and/or the emission lines are made longer than the rest of the wavelengths by an amount $\Delta\lambda$ (read delta lambda), which increases with the wavelength. The velocity of recession is given by the simple formula

$$v = \frac{\Delta\lambda}{\lambda} \times c$$

where λ is the normal rest wavelength and c is the speed of light. If the source and the observer were approaching each other, $\Delta\lambda$ would be negative and the velocity would be one of approach. As an example, let the rest $\lambda = 5000$ Å, $\Delta\lambda = +1$ Å, and $c = 300,000$ km/s (18,000 mi/s). Then substitution in the above formula gives a recession velocity of 59.9 km/s (+37.2 mi/s). In practice as many spectral lines are measured as possible, the velocity is determined from each $\Delta\lambda$, and the mean is taken for all values as the velocity of approach or recession between the earth and the source of light. The velocity obtained this way is called the **radial velocity** because only motion along the line of sight or along the radius can be determined by this means.

3.15   RADIO TELESCOPES

The newest and perhaps the most exciting field in astronomy is radio astronomy. The subject will be discussed in later chapters, but here some mention will be made of radio telescopes as instruments. Some of these telescopes look like gigantic bed springs, others are a cross with each arm a half-mile or more long; most of them resemble vastly oversized optical telescopes. See Figure 3.13.

Most radio telescopes are large for two reasons. First, the larger the surface area, the more radio energy it collects, and therefore, the stronger the signal. The second reason involves the characteristic of resolution discussed in Section 3.8. The radio waves studied have wavelengths ranging from about one centimeter to many meters. The minimum angle of resolution in Section 3.8 can be rewritten as $\theta = 69.9$ λ/d, where θ is in degrees and λ and d are in centimeters. One wavelength studied a great deal is the 21-centimeter (about 8-in) radiation from neutral atomic hydrogen in interstellar space. For a radio telescope with a diameter of 50 meters (155 feet), the resolution is not quite 0.3 degree. This value is vastly larger than the optical resolution for even a medium-sized optical telescope, but the wavelength used here is much longer (2.1 billion Å). High resolution is needed to obtain accurate positions for radio

FIGURE 3.13 A photograph of the 40-meter (131-foot) radio telescope at the Owens Valley Radio Observatory of the California Institute of Technology. (Photo courtesy Dr. Alan T. Moffet, Professor of Radio Astronomy.)

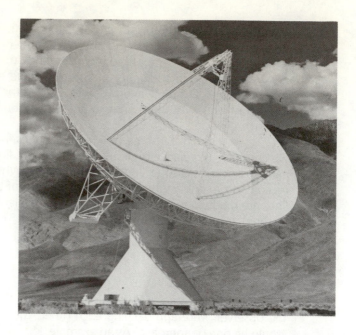

sources in the sky, to separate those that are close together, and to determine the structure of a source if its area is large. The largest radio telescope in the world is 305 m (1000 ft) in diameter and is located in Puerto Rico.

A radio telescope does not take a photograph. It measures the intensity of the radio waves coming from that part of the sky limited by the resolution of the instrument, as defined by the diameter of the instrument and the wavelength at which it is working. Some of the largest radio telescopes have been used as radar instruments in studies of the planets and the sun.

Recently developed techniques, using two radio telescopes several thousands of miles apart in the form of an interferometer, have permitted astronomers to attain resolutions of 0.001 arc-second. This permits detailed determination of the microstructure of some of the radio sources.

3.16 SYNCHROTRON RADIATION

In this chapter, the absorption and emission of radiation by atoms and molecules have been discussed, but there is one further type that should be discussed here for future use in this text. This is called **synchrotron radiation.** Both theoretically and experimentally, it can be shown that when a very high-speed electron moves through a magnetic field, not only does it change direction of motion, but also it radiates energy in all wavelengths with an intensity that increases with increasing wavelength. This behavior is contrary to the pattern of continuous radiation from a hot body in the long wavelength region (see Section 3.10). The spectrum extends from the short wavelength optical region into the infrared and on into the radio region, where it is most intense. Synchrotron radiation is usually observed to come from sources in which the gases are highly ionized, the free electrons are very energetic, and there are relatively strong magnetic fields. The magnetic field must be produced by currents of charged particles, and the high-speed electrons are produced either by some explosive process or by acceleration in the magnetic field itself. Later, when we discuss astronomical sources of radio energy, we shall have several occasions to refer to synchrotron radiation.

KEY TERMS

radio waves

infrared waves

visible light waves

ultraviolet

X-rays

gamma rays

angstrom

refraction

dispersion

index of refraction

spectrum

focus

focal length

chromatic aberration

spherical aberration

resolution

spectroscope

continuous spectrum

bright line spectrum

absorption spectrum

ionization

isotope

photon

excited state

Doppler effect

QUESTIONS

1. What would be the advantages to observing the sun, moon, planets, and stars with a telescope placed in an orbit around the earth above our atmosphere?

2. How could you determine the value of the astronomical unit in miles by making radial velocity observations of a star that lies in the plane of the earth's orbit?

3. What is the magnification of a telescope if the focal lengths of the objective and eyepiece are, respectively, 38 feet and ¼ inch?

4. Compare the relative merits of refracting and reflecting telescopes.

5. What would be the required diameter of the objective of a telescope that could just resolve the two components of a double star whose separation was 0".3?

6. What are some of the practical values of a knowledge of spectroscopy for industrial applications?

7. Consider the problems in building a telescope much larger than the 200-inch Hale reflector.

8. At what places in their orbits relative to the earth would you expect the radial velocities of the planets to be zero?

9. After reading Section 3.2, consider the view seen by a submerged fish when the surface water is quiet.

10. If you wished to hunt fish with a bow and arrow, how would you aim the arrow?

4 The First Solar Family: Planets, Satellites, and Asteroids

The solar system is a variegated assemblage of bodies extending many billions of miles in all directions from the sun. Were it not for the gravitational grip of the sun, all would be chaos, and in a short time the present system would be dispersed into space and anonymity. In this system, there are nine known planets, at least forty-three satellites and thousands of asteroids. The total solid volume of all this material is but a minute fraction of the space included in the solar system. We are left with the impression of a vast emptiness in which we might occasionally chance upon some tiny particle such as a planet.

In the 1930s, astronomical research was confined mostly to the stars, our galaxy, and the extragalactic system. Research on the solar system was considered to be a rather minor field. After World War II, radio astronomy and the exploration of the solar system with spacecraft began to open up new avenues of approach to the study of the individual members of the solar system and the interplanetary medium. The solar system is now a very active field for research and many unusual discoveries have been made, some of which will be discussed in this chapter. The work on our moon has already been discussed in Chapter 2.

4.1 THE SCALE OF THE SOLAR SYSTEM

No one, not even an astronomer, is capable of visualizing the enormous distances in the solar system. They are mere numbers. Imagine the sun, with its diameter of 864,000 miles, reduced to the diameter of a one-foot basketball. With all dimensions reduced on this scale, the earth would be a BB shot 109 feet from the sun. Jupiter would be a Ping-Pong ball 570 feet from the sun. Pluto, our most distant planet, would be smaller than the earth and 4300 feet from the sun. The distance to the nearest known star would be 5700 miles.

4.2 MEASURING DISTANCES IN THE SOLAR SYSTEM

Solar system distances are measured by a variety of means. Basically, this amounts to measuring the distance in miles from the earth to the sun. Some of the methods are quite complex, but here an attempt will be made to give the reader a feeling for the methods and to discuss one method in a simple way.

Let us remember Kepler's third law which states that $a^3 = P^2$, where a is in astronomical units (AU) and P is in years (Section 1.14). It is clear that if the period of revolution is known for a body revolving around the sun, one can use the formula for a determination of its semimajor axis in astronomical units. Let us consider the earth, where a is one AU and P is one year, and another planet whose period has been determined to be 1.1 years. Assume circular orbits for both planets. Squaring 1.1 and then taking the cube root of the result shows that the planet's orbit has a radius, a, of about 1.066 AU. In other words, when this planet is closest to the earth its distance is 0.066 AU. But how many miles is this? This would be easy to determine if you knew that 1 AU = 150 million km (93 million mi) (round numbers), but suppose you did not know that. Then you would have to measure by some means the distance in miles of the planet at closest approach. Many methods have been tried, including triangulation from widely separated points on the earth. The most recent and most successful method uses radar. It involves sending from a large radio telescope on the earth a very intense, very short time-length pulse of radio waves to the planet. The planet intercepts a small fraction of this radio energy and reflects it into space. Some of this energy is detected by the radio telescope, and the total time of travel of the radio waves from the earth to the planet and back to the earth is measured. The speed of light (that of radio waves) is well known, and the travel time can be measured with very great accuracy. Therefore in the example quoted above, one can determine how many miles there are in 0.066 AU and, as a result, how many miles there are in one astronomical unit. This distance of about 150 million kilometers is now known with an accuracy of about one kilometer.

This method has been used with great success using Venus as the reflector. Because the orbits of the earth and Venus are elliptical and not in the same plane, the problem is a little more complicated than the case cited above, but the principle is the same. Understand that the semimajor axes of the orbits of all the planets in AU, their spatial orientation, and the position of each planet in its orbit at any time are known with great accuracy. Therefore, the distance from the earth to any planet at any time can be determined in astronomical units and with very great accuracy. Venus has been used to determine the number of miles in one astronomical unit because it is frequently close to the earth. The minimum distance is about 42 million km (26 million mi). The measurement of the astronomical unit in miles has been a fundamental astronomical problem for a long time.

4.3 THE TWO SOLAR FAMILIES

Even a casual inspection of the solar system reveals that, except for the sun, all of its members fall into two distinct families—the family of the planets and asteroids and the family of the comets and meteors. Without exception, all the members of the first family revolve about the sun in direct (ccw) motion. The average orbital eccentricity is low compared to that of the second family. Among the planets, Pluto has the highest orbital eccentricity (0.25), and a few of the asteroids have even higher values. The average orbital inclination with respect to the plane of the earth's orbit is small, with Pluto again having the largest value (17 degrees). With the exception of Venus and Uranus, all the planets apparently rotate ccw on their axes. The direction is unknown for Pluto. Table 4.1 gives some data on the planets including the largest asteroid, Ceres, and Icarus.

In the second family, the comets and meteoroids, the situation is quite different. The average orbital eccentricity is much higher, and it may range from nearly zero to almost unity. The orbital inclination can range from zero to as much as 90 degrees, and the direction of revolution is often retrograde (cw). This classification strongly suggests that the two families have quite different origins.

4.4   THE PHASES AND ASPECTS OF MERCURY AND VENUS

Because both Mercury and Venus revolve about the sun inside the earth's orbit, they are called *inferior planets*. For the moment let us consider the positions and appearances of Venus alone.

Remember that the side of Venus facing toward the sun is illuminated, but that our view of Venus is from a point outside Venus' orbit. When Venus is between the earth and the sun, it is at **inferior conjunction** and we are looking at the dark side of Venus. We see nothing unless Venus passes closely between the two bodies. Then we see a rare **transit** of Venus across the sun's disc. The orbital planes of Mercury and Venus are not in the same plane as that of the earth's orbit. (See Table 4.1).

After inferior conjunction, the planet in its ccw motion around the sun moves to the right (or west) of the sun and becomes a morning star because it rises before sunrise. Shortly after inferior conjunction we see part of the bright side as a thin crescent which grows to quarter Venus when the angle sun-Venus-earth is about 90°. At this time the angle sun-earth-Venus (the **elongation**) is the greatest (about 46°), and Venus rises in the east about three hours before sunrise.

Continuing in its orbit Venus starts to move in toward the sun as seen from the earth and we see a larger fraction of the bright side. When it is on the other side of the sun from the earth, it is at **superior conjunction** (full Venus). It rises with the sun, but it is too close to the sun to be seen. After superior conjunction, Venus passes to the east side of the sun and becomes an evening star because it sets in the west after the sun. After a time it reaches **greatest eastern elongation** (again about 46°) and sets in the west about three hours after the sun sets. After this the fraction of the right side seen from the earth decreases to crescent shape and then to zero, when Venus is back at inferior conjunction. Near greatest elongation, east or west, Venus is a brilliant object and, with care, may even be seen by the naked eye in daylight. Because of the great distance change, the image of Venus is about six times larger near inferior conjunction than near superior conjunction.

Mercury exhibits the same aspects and phases except that it is usually fainter, closer to the sun, and harder to see even in twilight. Because of its eccentric and smaller orbit, the greatest elongations range between 18° and 28°.

TABLE 4.1 Planetary Data

Planet	Mean distance from sun (km, millions) (mi, millions)	(AU)	Orbital period	Orbital eccentricity	Inclination to ecliptic (degrees)	Equatorial diameter (km) (mi)	Mass (earth = 1)	Density (water = 1)	Rotation period
Mercury	58	0.39	88d	0.206	7.0	4880	0.05	5.4	58.6d
	36					3030			
Venus	108	0.72	225d	0.007	3.4	12,104	0.81	5.3	243.1d
	67					7,520			
Earth	150	1.00	365¼d	0.017	0.0	12,756	1.00	5.5	23h56m
	93					7,927			
Mars	228	1.52	687d	0.093	1.8	6,800	0.11	3.9	24h37m
	142					4,220			
Ceres	414	2.77	4.6y	0.079	10.6	980	0.0001	?	9h05m
	257					610			
Icarus	161	1.08	1.12y	0.827	23.0	1.4	?	?	2h16m
	100					0.8			
Jupiter	778	5.20	11.86y	0.048	1.3	142,800	317.9	1.3	9h52m
	484					88,730			
Saturn	1,429	9.55	29.46y	0.056	2.5	120,660	95.1	0.7	9h25m
	888					74,980			
Uranus	2,875	19.22	84.01y	0.047	0.8	50,800	14.5	1.2	23.9h
	1,787					31,570			
Neptune	4,500	30.11	164.79y	0.009	1.8	48,600	17.2	1.7	18.2h
	2,800					30,200			
Pluto	5,900	39.44	248.5y	0.250	17.2	3,000	0.002	1.0	6.39d
	3,666					1,860			

60

Mercury moves quickly around the sun in a rather elliptical orbit (see Table 4.1). For a long time it was thought that Mercury always kept the same face toward the sun, making its rotational period the same as its orbital period of 88 days. Radar data now show that the actual rotation period is the unusual value of 2/3 of its orbital period, or 58.6 days. (See Section 4.6.) The side toward the sun has a surface temperature in excess of 460°C (860°F); on the dark side it may get as cold as −150°C (−240°F). There is no evidence of an atmosphere. Until recently, only a few vague surface markings could be seen, but in 1974 the spacecraft *Mariner 10* was sent on a trip that passed Venus and Mercury in close encounters. *Mariner 10's* televised pictures of Mercury showed that the surface is covered by many impact craters, basins, fault escarpments, and ridges and has a strong resemblance to the lunar surface. See Figure 4.1. The terrain appears to be even more rugged than that of our moon. The basins are often large, walled impact craters in which the interior is quite flat and relatively smooth except for numerous small impact craters. The planet seems to have a weak permanent magnetic field.

Venus revolves about the sun in a nearly circular orbit inside the earth's orbit, and at inferior conjunction it comes within about 42,000,000 km (26,000,000 mi) of the earth. Hence, Venus comes closer to the earth than any other planet, but this advantage is lost because we are looking at the dark side. Even when the planet is seen at other phases, from the earth our eyes see an almost totally featureless

FIGURE 4.1 A photograph of heavily cratered terrain on Mercury taken by the *Mariner 10* spacecraft. Sunlight comes from the right. The width of the view is about 500 km (300 mi). (Photo courtesy of the National Aeronautics and Space Administration.)

whitish-yellow surface with only a few elusive markings that are too faint to use for the determination of the planet's rotation period. Photographs in ultraviolet light reveal only a few large light and dark markings which change with time. It was assumed (correctly) that Venus had an opaque, cloudy atmosphere which totally obscured any surface features. Sunlight absorbed in the upper atmosphere and reflected from the cloud layers was analyzed spectroscopically to show the overwhelming abundance of gaseous carbon dioxide.

Ultraviolet light photographs taken during the *Mariner 10* close encounter revealed a great deal of upper cloud detail (see Figure 4.2). A succession of photographs taken over several days showed an atmospheric circulation of hot gas rising from a point where the sun is overhead at the equator. Spiraling up toward both poles in about four days, the gas then descends in the polar regions and is presumably returned to the equator at lower levels. Because the cloud layer is opaque, the planet's rotation period cannot be determined from observations of surface markings. However, the radar technique described in Section 4.6 has shown that the rotation period is the long value of about 243.2 days. Most surprisingly, it has also shown that Venus' rotation is *clockwise* as opposed to that of the other planets, except Uranus. What is even more strange is that if the rotation period is 243.16 days, then at each inferior conjunction the same side of Venus faces toward the earth. This suggests that the gravitational attraction of the earth has somehow controlled both the direction and rate of rotation of the planet. In view of the small masses of both planets as compared with that of the sun, this suggestion is an unlikely one.

These results and more recent ones from probes sent to Venus by the United States and the USSR are summarized here. Venus has three cloud layers, the upper two of which are between altitudes of 70 to 50 km (43 to 30 mi) above the surface and are really haze layers adding little to the atmospheric opacity. The main obscuring cloud layer is only about 5 km thick at the altitude of about 48 km (30 mi), and the temperature in this layer is about 200°C (400°F). Below this layer there is more light haze, but from about 32 km (20 mi) down to the surface the atmosphere is clear. At the surface the temperature reaches the astonishingly high value of 480°C (900°F). This is an extreme case of the greenhouse effect of an atmosphere (see Section 4.7).

FIGURE 4.2 A close-encounter photograph of Venus taken by the *Mariner 10* spacecraft. One of the poles of rotation is at the upper right. Note how the clouds spiral from the equator to the poles in both hemispheres. (Photograph courtesy of the National Aeronautics and Space Administration.)

The atmospheric pressure at the surface is 90 times that on the earth at sea level. Winds up to 360 km/hr (225 mi/hr) are observed; they take about 4 days to circle the planet.

The temperatures in all layers are too high by far for the particles to be frozen carbon dioxide (dry ice), and the spectroscope shows the extreme scarcity of water in any of its phases. Very careful spectroscopic analyses of the atmospheric composition have just about confirmed the suggestion that the opaque clouds are really droplets of sulfuric acid. Sulfuric acid has a great affinity for water, thereby explaining the absence of free water vapor. The actual chemical processes which produce these particles involve a complex of reactions between free sulfur, oxygen, and water vapor.

Some sunlight does reach the surface, but because of the scattering of the shorter wavelengths by the cloud particles the whole landscape and the sky is a deep red and the sun would not be visible. A few years ago Russian probes reached the surface and showed that around each lander the area was covered by broken, angular-edged rocks, which suggest that wind erosion is not important and that surface wind speeds are low. The United States has placed a radar-equipped orbiter in motion around the planet. Radar beams from the orbiter to Venus and reflected back to the orbiter indicate that the surface is less rough than that of our moon and more like that of the earth's topography. The reasons for the tremendous differences between the atmospheres of Venus and the earth are a major scientific mystery.

4.6 THE RADAR DETERMINATION OF THE ROTATION PERIOD OF A PLANET

This technique sends a powerful beam of radar waves in a very narrow wavelength (frequency) range to a planet. If the planet is not rotating, there will be no Doppler shift (Section 3.14) in the waves reflected from any part of the planet's surface. There will be only the constant shift due to the relative motion between the planet and the earth. If the planet is rotating on an axis at right angles to the line to the earth (assuming no relative motion between the earth and the planet), the central regions moving at right angles to the earth will reflect the same wavelength received from the earth. The side moving away from the earth will reflect a longer wavelength, and the reverse will be true for the side approaching the earth. The reflected waves received back at the earth will show a broadened line instead of the sharp, narrow one transmitted. From this broadening, one can calculate the planet's rotation period. The inclination of the rotation axis and whether the rotation is cw or ccw can be determined from an extended series of observations.

4.7 THE GREENHOUSE EFFECT OF PLANETARY ATMOSPHERES

The earth's atmosphere is composed mainly of nitrogen, oxygen, carbon dioxide, and water vapor. This atmosphere is almost totally transparent to the solar radiation in the range from 3000 Å to 12,000 Å (the optical window) and it is in this range that the sun emits about 50 percent of its total radiation. If we disregard the almost negligible amount of ultraviolet radiation below 3000 Å, the radiation above 12,000 Å in the infrared is almost entirely blocked by carbon-dioxide and water-vapor absorption; a few scattered windows in the infrared have a reasonable transparency. Therefore, little of the sun's infrared energy reaches the surface but is instead absorbed in the high atmosphere far from the surface.

Some of the sun's energy which reaches the surface in the optical-window range is reflected back into space; the rest is absorbed by the surface and warms it. The air for a few thousand feet above the surface may be warmed by contact with the surface. The result is that the average surface temperature of the earth is about 300K in a round number (27°C, 80°F). Because of its temperature, it has a spectrum of continuous radiation which for 300K has a maximum at about 100,000 Å in the far infrared (see Section 3.10). Therefore, almost all of the earth's radiation is absorbed by the carbon dioxide and water vapor in the air close to the surface. This warms the air above what it would be if the absorbing gases were not present. At night, when there is no incoming solar radiation, the earth's surface continues to radiate, but due to the blocking effect of water vapor and carbon dioxide, the rate of escape of radiation is slow. Hence the temperature drop from sunset to sunrise the next morning is usually moderate. This phenomenon is often called the **greenhouse effect** because of the correspondence between the behavior of a greenhouse and that of the earth's atmosphere.

Mercury and the moon have no atmospheres. As a result there is no greenhouse effect and surface temperatures show a very large range between night and day. The other extreme is Venus, with a very dense carbon dioxide atmosphere which efficiently traps the incoming optical-window radiation and keeps the surface temperature at about 480°C (900°F). Mars is an in-between case, with a surface atmospheric pressure of only about 1 percent that of the earth, but the air is mainly carbon dioxide. Although the daily temperature variation at Mars' equator is about 110°C (200°F), the range would be more without its atmosphere.

4.8 MARS AND ITS MOONS

Mars has two moons, Phobos and Deimos. Phobos has an orbital period of 7^h39^m and a distance from the planet's center of only 9330 km (5800 mi). Deimos is 23,500 km (14,600 mi) from the Martian center and has an orbital period of 30^h18^m. Phobos revolves around Mars about three times in the Martian day of 24^h37^m; therefore, as seen from the planet's surface, it rises in the west and sets in the east—the only solar system satellite to do so. The respective diameters of Phobos and Deimos are about 29 and 21 km (18 and 13 mi). Photos from spacecraft close to the planet show that both satellites are heavily cratered and somewhat irregularly shaped. Phobos, but not Deimos, is marked by numerous scratches, many of which are parallel to one another but whose origin is unknown. Both satellites show that some of the smaller craters are partly obscured by a thin dust layer—probably from the planet itself.

Because of the inclination (24°) of its rotational axis, Mars' seasons are similar to our own; they are, however, almost twice as long. Even at closest approach to the earth, Mars' angular diameter is only 25 arc-seconds. Using a 40-inch telescope with a resolution of 0.1 arc-second, the smallest detail that could be resolved on Mars would be about 27 km (17 mi), and this only under the best of observing conditions. Under most conditions the telescopic view of Mars is disappointing. One can see that the planet's color is reddish, and large and dark areas are visible. When it is winter in one hemisphere, a large white polar cap can be easily seen; for both hemispheres this cap comes and goes with the seasons.

4.9 THE ATMOSPHERE AND TEMPERATURE OF MARS

Infrared photographs of Mars show the larger features quite clearly, but photographs in blue light are much less distinct and many show no features at all except the polar caps. This is evidence for a Martian atmosphere, because blue light waves are more easily scattered by gas molecules and dust particles than are infrared light waves; thus surface details are blurred in blue light.

Earlier from the earth and later from spacecraft sent to Mars, spectroscopic evidence showed the presence of the following gases (followed in each case by the volume percentage) in the planet's atmosphere: carbon dioxide (95), nitrogen (3), argon (1.5), oxygen (0.1), and water vapor (0.03), plus other gases in very small amounts. The surface atmospheric pressure on Mars averages a little less than 1 percent of that on the earth at sea level.

Surface temperatures are obtained by measuring the infrared radiation from the Martian surface and filtering out the reflected infrared solar component. The maximum equatorial temperature is about 27°C (80°F) around Martian noon, and it drops to about −84°C (−120°F) just before dawn the next morning. The average year-long Martian temperature is about −60°C (−76°F). Because the Martian orbital eccentricity is 0.093, the distance of Mars from the sun in the early southern Martian summer is 9.3 percent less than the mean orbital distance and that much more when it is the same season in the northern hemisphere. The result is that when it is early summer in the southern hemisphere that region receives some 45 percent more solar radiation than at the same season in the northern hemisphere; as a result the early southern summer is considerably warmer. At all times during the Martian year the mean daily surface temperature is below the freezing point of water so that all the ground at the surface is permanently frozen. The average depth of this permafrost layer over the whole planet is estimated to be about 2.5 km (1.6 mi). Again in the polar regions the winter temperatures are always below the freezing points of both water and carbon dioxide. Only for a brief part of a day and near the equator is the temperature high enough for water to melt; this does not occur in the polar regions.

4.10 LIFE ON MARS

The spirited controversy about life on Mars began in 1877 with Schiaparelli's discovery of a fuzzy network of lines which later came to be regarded as a network of canals. It was thought that these canals brought melted polar ice water to the dry equatorial regions during the Martian spring season in each hemisphere. Clearly, if canals exist, they must have been excavated by intelligent life forms with considerable engineering skill. As will be seen in the next section, the polar caps are frozen carbon dioxide. The spectroscopic evidence of little atmospheric water means that the surface is very dry and desertlike. Also, the almost total disappearance of the polar cap in summer shows that the ice layer is very thin. The reality of the system of lines interpreted as canals has now been disproven. The *Mariner* flights to Mars show that some of the stronger lines are either linear arrangements of craters or long, deep valleys. The low atmospheric surface pressure, the large diurnal temperature changes, the almost total lack of oxygen, and the small amount of water all conspire against existence anywhere of forms of life as highly developed as our own. Seasonal changes in the color of the surface suggest vegetation, but spectroscopic evidence rules out chlorophyll.

In late 1976 two spacecraft, the *Viking* soft-landers, touched down on the surface and began a number of experiments. One of these was to determine if the Martian soil contained any form of life based on the chemistry of carbon, as is our own. Several exhaustive trials were made on the surface material, but the results are rather inconclusive.

4.11 DETAILED SPACECRAFT OBSERVATIONS OF MARS

From 1965 through 1976 the United States sent several spacecraft to observe the planet at close range in order to obtain high-resolution photographs and other data. Each of the two *Viking* spacecraft consisted of an orbiter and a lander. When the craft got close to the planet the orbiter and lander separated. The former continued to orbit the planet while taking photographs and making measurements of the planetary surface temperatures and the altitudes of surface features. Both landers reached the surface without damage and made an enormous number of observations. The orbiters took a total of 55,000 photographs before losing their capability to orient their cameras. See Figure 4.3. One of the landers is still operating and is expected to continue doing so until about 1990. The following paragraphs are a brief summary of some of the most important conclusions reached about Mars from the *Viking* spacecraft.

FIGURE 4.3 A composite photograph of Mars made from four photos taken by the *Viking 1* orbiter. Observe the large plain somewhat to the right of and above center and also the very large number of impact craters and mountains. The horizon is to the right; just above it are two layers of thin clouds probably made up of carbon dioxide crystals. Sunlight comes from above. (Photograph courtesy of the National Aeronautics and Space Administration.)

Not surprisingly, spectroscopic observations showed that the polar caps consist mainly of frozen carbon dioxide (dry ice) and most probably a considerable amount of water ice. The polar caps form in the winter in their respective hemispheres, disappear as spring and summer approach, and reform at the opposite pole. In midwinter the cap may be as large as 1900 km (1200 mi) across. The dry ice condenses from atmospheric carbon dioxide as winter approaches for a particular pole and then evaporates (sublimes) as summer nears. As seen from the earth the cap seems to vanish completely, but the high-resolution photos taken by the orbiters show a residual white mass which is water ice. This ice occurs in what appear to be seasonal layers separated by layers of dark dust; at this time very little, if any, of the ice sublimes except in the summer season and even then the amount is small.

Before the close-up photos of Mars were made, astronomers expected that most of the features would be like those on our moon. While it is true that the photos showed many craters, a few seas, and other features like those found on our moon, there is a great deal of difference. The most obvious type of terrain is characterized by volcanoes, of which there are many. The largest is Olympus Mons, which is about 550 km (340 mi) in diameter at the base and 27 km (17 mi) high; at the top there is a caldera 90 km (55 mi) across. Olympus Mons is the largest known volcano in the solar system. There are a great many other volcanoes of all ages, from as old as perhaps 4 billion years to perhaps less than a million years or even less. In some areas it is clear that great masses of lava have covered older features. Large areas of the planet are peppered by what are obviously impact craters like those on our moon and just as numerous in impact craters per square kilometer as on the moon. The cratering is almost entirely of ancient origin (4.5-4.0 billion years ago) as in the case of our moon; other more recent craters are much fewer in number.

The chaotic Martian terrain covers large areas of the planet and consists of ridges and valleys arranged in a randomly oriented pattern. This terrain was probably formed by internal movements which cracked and distorted the surface some time after the original solid crust was formed. There are no such features on our moon. The featureless terrain consists of large areas, some several hundreds of kilometers across, in which there appears to be almost no detail except for a rare crater of small size. Generally, these areas appear smooth and are somewhat depressed below the average level of the Martian surface. They resemble the lunar maria and may be large impact features; some of them have been later modified by other processes such as lava flows.

Of particular interest on Mars are the numerous large chasms. The largest of these is in the equatorial region. It has a length of more than 1600 km (1000 mi), an average width of 100 km (60 mi) and a depth of about 7 km (4 mi). These chasms resemble the rift valleys found on our earth and most probably are the result of faulting.

One of the most interesting and perplexing of the Martian surface features is shown in Figure 4.4. The total length of the feature from left to right is about 250 km (150 mi). The feature is dotted by a number of impact craters of different sizes. The most conspicuous features are the many channels which geologists recognize as the result of massive erosion caused by the rapid flow of enormous volumes of water. No one seriously proposes that the channels are the result of lava flows or that they are collapsed lava tubes. The channels are thought to represent one or more catastrophic events resulting in the release of immense quantities of water. But where did the

FIGURE 4.4 A photo of Mars taken by the *Viking 1* orbiter. The area is about 300 km (180 mi) from left to right and the sun is off to the right. Note the sinuous rille near the top where the ground slopes from left to right. The main feature is the collapsed area to the right of center, with channels leading off to the right and downhill. The collapsed area was probably caused by the melting of permafrost, and the channels were carved by the sudden release of a flood of water. (Photograph courtesy of the National Aeronautics and Space Administration.)

water come from? Clearly there is no flowing water now on Mars because it is far too cold. A slow warming climatic trend will not do. If an increase in the interior temperature took place and heat began to flow to the surface, the deeper regions of the presumed permafrost would melt but the water would be prevented from escaping to the surface by the hard upper layers of permafrost. When finally the pressure became too great the water might burst through to the surface and scour out the channels. There is evidence that in some areas highland regions have collapsed, allowing the start of the channels, some of which may be thousands of kilometers long. The source of the water and the mechanism of its release is the greatest enigma concerning the geological development of Mars.

In 1976 the United States soft-landed two capsules on the Martian surface at a considerable distance apart. Both landers made many meteorological measurements, took thousands of photos of the nearby Martian surface, and looked for evidence of life. One of the photos is shown in Figure 4.5. In this area the surface is strewn with boulders as large as 2-3 feet in diameter. Probably these boulders are the ejecta from impact events. Some of the larger rocks show small sand dunes on the side opposite to the wind direction, but the sharpness of the boulder edges suggests that the dust is very fine and is not a good eroding agent even at high speeds. Spectroscopic evidence shows that the dust has a high silica content. The rocks are red at both lander sites, probably indicating the presence in the rocks of iron-bearing minerals. In other places on Mars the orbiters photographed large areas 160 km (100 mi) across covered by immense sand (or dust) ridges. The dust storms mentioned earlier can transport large quantities of dust which are deposited in the depressed areas—a possible reason why part of the terrain is featureless.

The development of Mars seems to be intermediate between that of the earth,

FIGURE 4.5 Photograph taken from the surface of Mars by the *Viking 2* lander in 1979. The large rocks in the foreground are 2-3 feet across, and the distant horizon is about one mile away. The white coating on the rocks and soil is a thin coating of water ice or frost. Most likely these rocks are the ejecta from impact craters. (Photograph courtesy of the National Aeronautics and Space Administration.)

with its heavy atmosphere and abundance of water, and the moon, with no atmosphere or water. The Martian vulcanism and the chaotic terrain must be the results of internal motion in a fairly hot interior. The magnitude of this internal motion is also probably intermediate between that of the earth, which exhibits considerable internal motion, and that of the moon, which seems not to have experienced this behavior for at least three billion years. It will be important to place seismographs on Mars as was done on our moon. One of the *Viking* landers did have a seismograph, but it failed to operate. While continental drift is now well established on our earth, there is no evidence that it has ever operated on Mars. Since Mars has a mass intermediate between that of the moon and the earth, it is generally believed that the internal structure and movement, as well as the temperature, are directly related to the mass of the body.

4.12   BODE'S LAW—THE ASTEROIDS

In 1772 the German astronomer Bode called attention to an interesting numerical progression relating the distances of the planets from the sun to their numerical order. Write down the numbers 0, 3, 6, 12, and so on, doubling the number each time. Then to each of these numbers add 4 and divide the sum by 10. Reference to Table 4.1 will show that these numbers give rather closely the distances in AU of the planets Mercury through Saturn, the last one being the most distant planet known in Bode's time. However, there was no planet at 2.8 AU as there should have been if the progression were to have meaning.

On the first night of the 19th century, Giuseppe Piazzi, a Sicilian monk, discovered a relatively faint object which changed its position with respect to the stars on succeeding nights. It was soon shown to be a planet with a semimajor axis of 2.8 AU, in seeming confirmation of Bode's law. It was named Ceres. See Table 4.1. As time went on, the even more startling discovery was made that many more of these objects exist in the space between Mars and Jupiter and that the average of their semimajor axes is close to 2.8 AU. Does this imply that Bode's law is some kind of fundamental relation? The extension of the progression beyond Saturn predicts the distance of Uranus rather well but not that of either Neptune or Pluto. It is possible that the progression is no more than a fortuitous relation.

The objects being discussed are most often called **asteroids**. Several thousand have been discovered. The great majority have semimajor axes between about 2.1 and 3.3 AU, averaging close to 2.8 AU. *All* revolve in direct motion—there are no exceptions. Their orbital periods range from about 3.0 to about 6.0 years. For the nine major planets the average orbital eccentricity is 0.08, but for the asteroids it ranges from about 0.1 to about 0.3. Again for the planets the average orbital inclination is about 4.0, while that of the asteroids ranges from $0°0.$ upward to about $30°$ A few of the asteroids have highly elliptical orbits. One of these is Icarus (see Table 4.1), for which the eccentricity is so high and the semimajor axis so small (1.08 AU) that at perihelion it comes closer to the sun than does the earth. Asteroids of this type have a finite chance of striking the earth over long time periods.

Ceres, the first asteroid to be discovered (and the largest), has a diameter of 770 km (480 mi). The diameters of only a few asteroids can be determined by direct observation; for the rest the angular diameters are too small to be resolved. The observable asteroids have diameters as small as about one-half mile. These diameters are calculated from the brightness of the reflected sunlight and an estimate of the reflectivity (albedo) of the asteroid's surface. The fact that many show periodic brightness fluctuations must mean that their shapes are irregular and that almost all the asteroids resulted from some collisional event. Well-measured rotation periods range from 2.3^h to 18.8^h, although some are most certainly much longer. There must be a lower limit because, if the rotation period is too short, the asteroid would fragment from an excess of centrifugal force over that of self gravity and the adhesive strength of the asteroid material. It has been estimated that there are several tens of thousands of asteroids, but that their combined mass cannot be greater than about 1/1000 that of the earth, or somewhat less than one-tenth that of our moon. Their origin will be discussed in a later section.

Recent work indicates that there are several kinds of asteroids with quite different albedos ranging from about 5 to about 15 percent. Those of lower albedo are thought to contain black, tarry, carbonaceous compounds like a few meteorites found on earth, while those with higher albedo are stony or silicate bodies. There are a few cases in which asteroids might possibly be binary objects revolving about each other, but that is yet to be proven. Accurate orbits have been determined for about 2700 asteroids.

4.13 THE ESCAPE OF ATMOSPHERES

We have seen that Venus and the earth have substantial atmospheres, Mars has a very thin one, and our moon and Mercury have no appreciable atmosphere. We will see that Jupiter, Saturn, Uranus, and Neptune have massive atmospheres. To under-

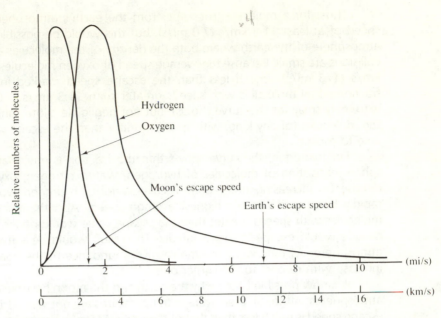

FIGURE 4.6 Speed distributions for the molecules of hydrogen and oxygen in a gas at a temperature of about 300K.

stand this situation, we shall need to explain what is meant by the **escape speed** from a planet and the speed distribution of the molecules in a gas.

Neglecting atmospheric friction, the escape speed from a planet's surface is that speed with which a body would need to be projected from the surface in order that its speed would approach zero as the distance from the earth approached infinity. Clearly the body would never return. For any planet this escape speed depends upon the square root of the planet's mass divided by its radius, i.e., the larger the mass and the smaller the radius, the greater the escape speed. This also means that the greater the pull of gravity at the planet's surface the greater the escape speed. In kilometers/second and miles/second, respectively, this value for a number of objects is: the earth 11.3 (7.0), our moon 2.4 (1.5), Mercury 4.2 (2.6), Venus 10.0 (6.2), Mars 5.2 (3.2), Jupiter 56 (35), and the sun 620 (384).

In any gas the average speed of the molecules depends directly upon the absolute (Kelvin scale) temperature. Figure 4.6 is a plot for hydrogen and oxygen molecules of the relative number of molecules at each speed at 300K, which is about room temperature. At zero speed the number is zero, but as the speed increases to the right on the plot, the number of molecules at each speed increases rapidly to a maximum at some value and then tapers off more slowly with larger and larger values of the speed; thus the number of molecules beyond some increasingly high speed becomes slowly smaller at high speeds. Further, at any given temperature, the average speed decreases as the molecular mass increases. At 300K the average speed of oxygen molecules is about 1/2 km/s (1/3 mi/s); for hydrogen molecules, it is four times greater at the same temperature.

THE ESCAPE OF ATMOSPHERES

Thus, for a molecule to escape from the earth's atmosphere it must have a speed of at least 11.3 km/s (7.0 mi/s), but this would be possible only in the high atmosphere of the earth where both the density of the molecules and the chance of collisions are small. Because the average speed of oxygen molecules is only about 1/2 km/s (1/3 mi/s)—much less than the escape speed marked in Figure 4.6—the fraction of all molecules with speeds greater than 11.3 km/s (7.0 mi/s) would be extremely small (as the curve shows), but not negligible. Note: only that fraction of the molecules (of any kind) with speeds greater than the escape value is potentially able to escape.

For hydrogen, the curve shows that the fraction is much larger. In an atmosphere consisting of molecules of hydrogen, water, nitrogen, oxygen, and carbon dioxide (in increasing order of molecular weight), hydrogen would escape most rapidly because it has the highest average speed. Also the fraction of hydrogen molecules with speeds greater than the escape value would be the greatest. Carbon dioxide would escape the least rapidly. The result would be a fractionation of the atmosphere. As time went on, the relative abundance of the heavier gases would increase with respect to the lighter ones.

It is now possible to see why the earth and the moon have such widely different atmospheres. Although their average temperatures are not very different, the moon's escape speed is much less than that of the earth. Mercury has no atmosphere because of its relatively low escape speed and its high temperature, which increase the escape rate. Mars has an intermediate atmosphere and an average temperature less than that of the earth, but note that heavy carbon dioxide is the most abundant gas. Fractionation seems to have operated here. The situation on Venus is not far different from that on the earth. The somewhat higher average temperature and slightly lower escape speed should promote escape, but the almost exclusive presence of carbon dioxide and possibly even heavier molecules may account for the dense atmosphere. Jupiter has a very massive and extensive atmosphere composed almost entirely of hydrogen and helium; there are also some heavy molecules. Both its high escape speed and its low temperature retard escape to a very great extent, so it is quite possible that its present atmosphere may be much the same as it was a few billion years ago. The same factors affect the atmospheres of Saturn, Uranus, and Neptune. One must not assume that the factors discussed here are the only ones that determine the composition and density of a planet's atmosphere, but they do contribute to our understanding of the present state of planetary atmospheres. Chemical processes must also be a factor.

4.14 THE GIANT PLANETS

Mercury, Venus, Earth, and Mars, and also our moon, are not far from being of the same diameter, but they are very much smaller than Jupiter, Saturn, Uranus, and Neptune. Pluto, too, is much smaller; until recently it was thought to be a possible escaped satellite of Neptune and not one of the original planets. But now that Pluto is known to have a satellite (1978), our view of the planet's origin is uncertain (see Section 4.20). The first four planets and our moon have a mean density of about 4.7 times that of water and all are rocky objects with a liquid or semiliquid core. The giant

Giants

planets are much more massive, but their mean densities are much lower. Each consists mainly of hydrogen and some helium with an admixture of a few other gases in relatively small amounts. It is probable that these planets have a very small rocky core around which is a very large shell of frozen hydrogen under very high pressure. Outside of this is a thinner shell of liquid gases and lastly an outer atmosphere. Spectroscopic studies of this atmosphere show the presence of hydrogen, ammonia, and methane. Jupiter is the most massive of these four planets, with about 1/1000 of the sun's mass: its mass is equal to that of the other eight planets combined. Our information has been greatly increased by four spacecraft: *Pioneers 10* and *11* in, respectively, late 1973 and late 1974, and *Voyagers 1* and *2* in 1979. All spacecraft made many measurements of temperatures of the planet and its satellites, made spectroscopic studies of the planet, and took thousands of photographs. These flights were remarkable achievements of American space technology. The major results will be discussed in the following sections.

4.15 JUPITER

Jupiter is truly the king of the planets. Its diameter and mass are the largest among the planets, yet its mean density is only 1.35 times that of water. From a telescope on earth, we observe dark and light bands parallel to the equator. These are cloud features and not solid surface markings, because the band widths change with time and the rotation period is not the same for all bands. The data from the four spacecraft show that Jupiter's dark and bright bands are the result of a convective circulation driven by the excess heat energy produced within the planet over that which it receives from the sun. The bright zones are rising clouds of gas whose tops are 15–20 km (9–12 mi) higher than the descending, somewhat warmer, dark cloud belts. Color and brightness of the bands may result from the dissociation and recombination of compounds to other compounds with changes in temperature.

Figure 4.7 is a *Voyager 1* photograph of the planet taken from a distance of 4 million km (2.4 million mi). It is far superior to any photo taken with the largest earth-based telescopes. The cloud bands and the Great Red Spot (below right of center) are conspicuous. The spot was probably first seen about 1660 and it is still visible, although its color, size, and shape have changed a good deal. The spacecraft show that the Great Red Spot is a long-lasting violent, cyclonic storm of unknown origin. On the earth most cyclonic storms persist for not longer than a week but here is one over 300 years old. The reason for the persistence of the Great Red Spot is a mystery. At this time the spot is about 40,000 km (25,000 mi) long. The same photo shows numerous other spots that are storms on a much smaller scale, but most of these last only a few months.

The infrared spectrum of Jupiter shows the absorption in reflected sunlight of ammonia, methane, and small amounts of ethane (C_2H_6), acetylene (C_2H_2), water (H_2O), phosphine (PH_3), and hydrogen cyanide (HCN). As discussed in the introduction (Section 4.14), some evidence for the internal constitution of the planet comes from the planet's *oblateness* of 1/15, which is the fraction by which the polar diameter differs from the equatorial diameter. The planet's high equatorial rotational velocity of 43,500 km/hr (27,000 mi/hr) causes a very high centrifugal force. This force and

FIGURE 4.7 Jupiter photographed by *Voyager 1*. The Great Red Spot is just below the center. Note the extreme turbulence between the upper bright band and the narrower dark one just below. Many other bright and dark oval storms can be seen in both hemispheres. The satellite Ganymede can be seen to the lower left of the planet. (Photograph courtesy of the National Aeronautics and Space Administration.)

the large oblateness show that the planet as a whole has a low rigidity. For the earth, with an equatorial rotational velocity of only 1700 km/h (1060 mi/h) and a composition mostly of rock, the oblateness is only 1/298.

Because of its very large mass and gravitational attraction, Jupiter disturbs to some degree the motion of all the planets, depending on their individual masses and their distances from Jupiter at a given time. The effect can be very large for low-mass asteroids and comets which come close to the planet. See Section 5.3.

Jupiter has a magnetic field about ten times stronger than that of the earth. Electrons are known to be trapped in the earth's magnetic field and spiraling about the lines of force. There are many more electrons trapped in the same way in the Jovian magnetic field. These electrons produce copious amounts of synchrotron radiation (see Section 3.16) and thereby make Jupiter a fairly strong source of radio noise.

The total energy in all wavelengths radiated by Jupiter is about twice that which the planet receives from the sun. One explanation for this is that Jupiter is slowly contracting and that the loss of gravitational energy is converted into radiant energy.

Another possibility is that this excess heat is a pale reminder of a much higher planetary temperature when the planet was being formed a few billion years ago. The present average cloud surface temperature is about −140°C (133K, −220°F).

One of the most striking discoveries made by *Voyager 1* took place as the spacecraft passed through the planet's equatorial plane and photographed the thin edge of a ring around the planet. The ring is less than 30 km (18 mi) thick and about 8000 km (5000 mi) wide. The ring's center is about 126,000 km (78,000 mi) from the planet's center and just inside the orbit of Metis, Jupiter's innermost satellite, which has an orbital radius of 128,000 km (79,500 mi). The composition of the ring material is unknown, but is probably fine dust. The amount of ring material must be small because it could only be seen when the spacecraft was close to the ring plane. The ring cannot be seen from the earth because of the planet's glare.

4.16 SATURN

The telescopic view of this planet is one of the most beautiful sights that astronomy has to offer (Figure 4.8). The surface of the planet resembles that of Jupiter except that the cloud bands are less distinct. The atmospheric composition is much the same except that the ammonia absorption lines are weaker. Because the cloud surface temperature is only about −170°C (100K, −270°F), it is assumed that much of the ammonia is frozen out of the atmosphere. It has been suggested that the low visibility of the cloud bands is the result of a haze of tiny ammonia crystals in the high atmosphere. The planet's mean density is about 0.7 times that of water, which must mean that a very large fraction of the planet's volume is gaseous. See Table 4.1. Recent evidence shows that Saturn, like Jupiter, radiates about twice as much energy as it receives from the sun, and for the same reason. Saturn has a magnetic field about 1/10 the strength of Jupiter's field.

The most distinctive telescopic features of Saturn are the two rings, the outer-

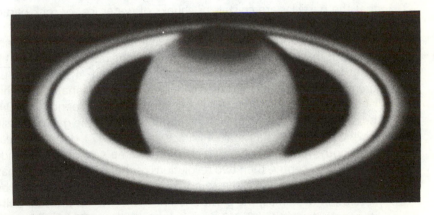

FIGURE 4.8 A photograph of Saturn taken by the 100-inch reflector on Mount Wilson. Note that the band system is less distinct than that of Jupiter. The two outer bright rings are clearly visible, as is the separation between them, called Cassini's division. (Photograph courtesy of the Mount Wilson and Las Campanas Observatories, Carnegie Institution of Washington.)

most A ring and the wider, brighter B ring. Careful observation in a good sky can show an even smaller, fainter C ring. Some observers suggested an innermost D ring. All have been confirmed. There are three more rings (E, F, and G) outside the A ring, but these are quite faint. The A ring has an outer radius of 137,000 km (84,700 mi) and an inner radius of 122,000 km (75,600 mi). The gap between the A and B rings, called Cassini's division, has a width of 4500 km (2800 mi). The bright B ring is 25,500 km (15,800 mi) wide. The innermost edge of the D ring is only 6600 km (4100 mi) from the planet's cloud tops. For lack of space the outer, very faint E, F, and G rings will not be discussed here.

The rings must be very thin (perhaps less than 1–2 km thick) because when both the sun and the earth are in the ring plane, the rings disappear. They cannot be solid sheets because when the slit of a spectoscope is placed across a ring's diameter, the spectroscopic Doppler shift shows that the inner edge of the ring has a larger orbital velocity about the planet than does the outer edge; the difference in velocity agrees with Kepler's third law at those distances. Therefore, the rings are not solid sheets. If they were, the outer edge would have the greater orbital velocity. The rings must therefore be made up of particles, each of which acts as an individual satellite. The present evidence for particle sizes suggests a range from the finest dust up to objects a few meters in diameter. The fact that the rings are not opaque is apparent when the planet and its rings pass between the earth and a fairly bright star—the star is dimly visible through the A and B rings and only slightly fainter when seen through the inner C and D rings.

It has been suggested that the ring particles are fragments produced by the tidally disruptive action of Saturn on a satellite that came too close to the planet. Every planet has a limiting distance called **Roche's limit**, inside of which a satellite would be torn to pieces by the tidal forces of the planet. Saturn's larger satellites are all outside the planet's Roche limit, but some of the smallest are not. Another theory is that the rings are material left over from Saturn's formation. Because this material was inside the Roche limit, it was prevented from collecting together to form a satellite. Recent observations of the sunlight reflected from the ring particles indicate that the particles are crystals or chunks of water ice, or rock particles coated with a layer of water ice.

Observations from the earth gave little evidence that the rings had any structure, but the *Voyager* observations revealed the unexpected—that the rings consisted of at least a thousand narrow ringlets. See Figure 4.9. This observation came as a great surprise and at this time there is no good explanation for the great multiplicity of ringlets. Consider the satellite Mimas, which has a period of 0.942 day. It can easily be shown that a particle revolving about Saturn in the gap of Cassini's division would have a period just half that of Mimas, that is, 0.471 day. This means that if we start with both Mimas and the particle on the same side of Saturn and along a line from the planet's center, then two revolutions of that particle later Mimas, the particle, and Saturn will again have the same configuration. The result will be that, after each revolution of Mimas, the particle will be subject to an unusual pull from Mimas, with its constant period of 0.942 day; in due time the particle will be pulled out of the gap of Cassini's division. Here the ratio of the particle period to that of Mimas is 1:2. For the next two moons (counting out from the planet), the ratio of the particle's period in the Cassini gap to that of the satellite is, respectively, 1:3 and 1:4. For a particle in the boundary between rings B and C the period is 1/3 that of Mimas. Other moons

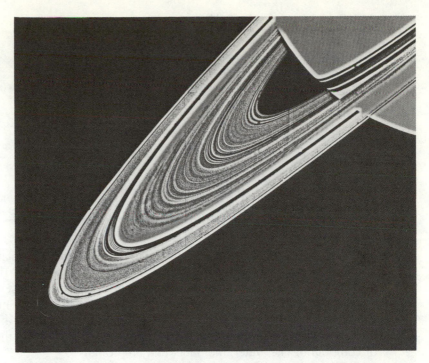

FIGURE 4.9 Details of the ring system of Saturn. This photo was taken from a distance of 8 million km (5 million mi). The original photograph shows about 95 individual ringlets, but on those taken at much closer distances it is possible to resolve as many as 1000 ringlets. The group of four rings about 1/4 of the distance in from the outside are in Cassini's division, showing that, as seen from the earth, these divisions are not empty. The few black dots on this photo were superimposed for the purposes of measurements. (Photograph courtesy of the National Aeronautics and Space Administration.)

can be shown to produce other gaps, but there is no way in which this process could produce as many as 1000 rings. This enigma is keeping the theorists very busy.

4.17 THE SATELLITES OF JUPITER

The satellite systems of both Jupiter and Saturn are remarkable subsolar systems themselves. Jupiter has at least 16 satellites, but there may be more too small and faint to be seen by present methods. Their distances from the planet's center range from 128,000 km (79,500 mi) to 23,700,000 km (14,700,000 mi) and their diameters from a few kilometers to 5280 km (3280 mi) for Ganymede. Their revolution periods about the planet range from 7 hours to 758 days. Data concerning the four largest satellites are given in Table 4.2. These satellites were discovered by the Italian astronomer Galileo in 1610 while he was studying the planet with the first telescope ever used for astronomical purposes. Note that Callisto has nearly the same diameter as Mercury and that Ganymede is even larger. Observe that the mean density decreases from 3.6 for Io to 1.8 for Callisto, the most distant of the four. The value for

TABLE 4.2 Data on the Galilean Satellites

Name	Distance from planet's center (km, thousands) (mi, thousands)	Orbital period	Diameter (km) (mi)	Mean density (water = 1)
Io	413 260	1ᵈ18ʰ	3632 2260	3.6
Europa	360 420	3ᵈ13ʰ	3100 1930	3.0
Ganymede	1070 665	7ᵈ04ʰ	5250 3260	1.9
Callisto	1882 1170	16ᵈ17ʰ	4850 3010	1.8

our moon is 3.3. This change in density with distance from the planet may be due to Jupiter having had a fairly high temperature at some past time. This would make Io much warmer than Callisto, with the result that Io would have lost most of its volatiles and would be composed only of rock-type minerals of a much higher density than water. At the other extreme Callisto would not have been much affected by Jupiter's heat; hence Callisto ended, as it is now, as a mixture of rock and the ices of a number of compounds like water, methane, and ammonia.

The outermost four Jovian satellites, at distances of close to 22.5 million km (14 million mi), all revolve in the **retrograde** direction. These four and perhaps some of the other, smaller satellites may be captured asteroids.

As seen from the earth the Galilean satellites are eclipsed periodically by Jupiter itself, or they may pass into the planet's shadow and disappear. The fact that the disappearance is not sudden but gradual shows that the edge of Jupiter is not sharp and that it has an atmosphere of gradually increasing density with depth. These four satellites have orbital planes with close to zero inclinations with respect to Jupiter's equatorial plane. When the earth passes through this plane there are usually several occasions in which two satellites eclipse each other. The duration of these eclipses gives good information about satellite diameters.

The *Voyager 1* and *2* flybys provided a flood of new information about Jupiter, but even more about the masses, diameters, densities, and surface characteristics of the satellites. One can imagine the surprise of the investigators when they found that Io had at least eight active volcanoes. The whole satellite appears to be very hot. This must be the result of Io's changing tidal distortion, produced by Jupiter and the other Galilean satellites, and resulting in tidal friction and heat. The volcanoes spew out dust, gas, and lava. Io's spectrum shows no evidence of water vapor, and the multicolored surface must consist of cooled lava flows. It is known also that the whole orbit of Io is outlined by a cloud of sodium and sulfur gas atoms, with Io at the center of the thickest part. Because Io and the other Galilean satellites have escape speeds of 2.7 km/s (1.7 mi/s) or less, none can retain a permanent atmosphere. No doubt the gases in the cloud ring escaped from Io itself.

Europa shows the least contrast of the four satellites. There are only a few impact craters of small size, but there are a number of linear markings about 100 km (60 mi) wide and at least 1600 km (1000 mi) long. These could be fissures or merely streaks produced by some unknown process.

Ganymede has only a few small craters except for one that is 800 km (500 mi) across and from which radiate what appear to be streams of ejecta. The topographical relief is low, probably because the rock-ice surface mixture is not rigid enough to support large altitude differences. There are long prominent scars, which may be fault lines, across the surface.

Callisto (Figure 4.10) shows many impact craters, but they are shallow like those of Ganymede because the surface is not very rigid (as for Ganymede). The very large impact crater near the top of the photo shows the roughly circular impact ridges characteristic of a surface material of relatively low rigidity. Europa, Ganymede, and Callisto show in the reflected sunlight the absorption spectrum of water ice which undoubtedly is more than just a superficial layer of water frost.

Future space missions to Jupiter (and Saturn) might be able to place spacecraft in orbit about the planet for long enough to make more extensive series of photographs and to make measurements of temperatures and magnetic fields. Such observations would enable astronomers to look for changes on the planet and obtain closer looks at the satellites. The prospect is a tempting one and is certain to yield important and exciting results.

4.18   THE SATELLITES OF SATURN

At present Saturn is known to have seventeen satellites, several of which were discovered by the *Voyager* spacecraft. Their diameters range from 16 km (10 mi) to 5150 km (3200 mi) for Titan, the largest. Distances from the planet's center range from 138,000 km (86,000 mi), just barely outside the A ring, to nearly 13 million km

FIGURE 4.10 Jupiter's satellite Callisto, showing the very large number of impact craters in a surface made up mostly of ices and rock material. Near the top observe the very large impact crater surrounded by ripple rings extending out from the center for several hundred kilometers. (Photograph courtesy of the National Aeronautics and Space Administration.)

(8 million mi) for outermost Phoebe. The corresponding revolution periods are 14.1 hours and 550 days. Phoebe is the only one of Saturn's satellites with retrograde motion.

The largest of Saturn's satellites is Titan, with a diameter of 5150 km (3200 mi), a distance from the planet's center of about 1.22 million km (760 thousand mi), and a revolution period of nearly 16 days. Titan (and possibly Triton, a satellite of Neptune) are the only satellites in the solar system with an atmosphere. That of Titan is composed mainly of nitrogen and some methane. Because Titan is a very cold object it may just be that at the bottom of its rather dense atmosphere is a solid surface of ice covered in places with liquid nitrogen oceans. The observed surface of the satellite's atmosphere shows almost no cloud markings.

Smaller than Titan but larger than the minor satellites are seven with diameters from about 400 to about 1500 km (250 to 900 mi). From measures of mass and diameter the densities have been determined to be low and indicate a high ice content; some may be nearly pure ice with little rock. These satellites differ considerably in appearance. Some are cratered but others are nearly smooth. Enceladus (Figure 4.11) has the highest albedo of any object in the solar system. Dione shows a dark surface with randomly placed wispy bright streaks. Mimas exhibits many impact craters and one huge crater that is about one-quarter (100 km) the diameter of the satellite. Iapetus, like our own moon, always keeps the same face toward the planet,

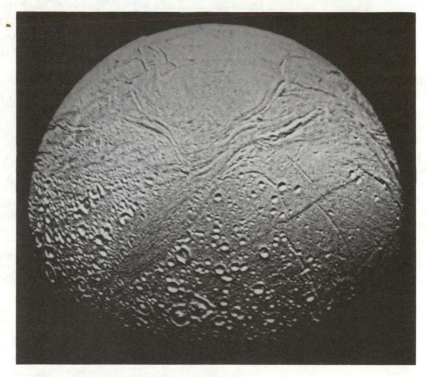

FIGURE 4.11 This is a high-resolution photograph of Enceladus, one of Saturn's satellites. Note the heavily impact-cratered surface and the numerous wavy channels, whose origin is unknown. (Photograph courtesy of the National Aeronautics and Space Administration.)

THE FIRST SOLAR FAMILY: PLANETS, SATELLITES, AND ASTEROIDS

but in its orbital motion the forward-facing side is much darker than the opposite trailing hemisphere. It is clear that each satellite has its own distinctive history of formation.

4.19 URANUS AND NEPTUNE

The diameters and rotation periods of Uranus and Neptune are somewhat uncertain (see Table 4.1). Uranus has five known satellites (see Figure 4.12) and Neptune has two. Those of Uranus have nearly circular orbits, but for the two of Neptune, Triton, in the smaller orbit, has an eccentricity of zero, while Nereid, in a much larger orbit, has an eccentricity of 0.75, the largest for any satellite or planet in the solar system. The surface temperature of Uranus is about 95K and that of Neptune is about 50K. *Voyager 2* will pass Uranus in January 1986 and Neptune in August 1989. Triton, with a diameter of 3200 km (2000 mi), appears to have a low-density atmosphere of methane.

Uranus is unusual in that if one is to regard its axial rotation as counterclock-

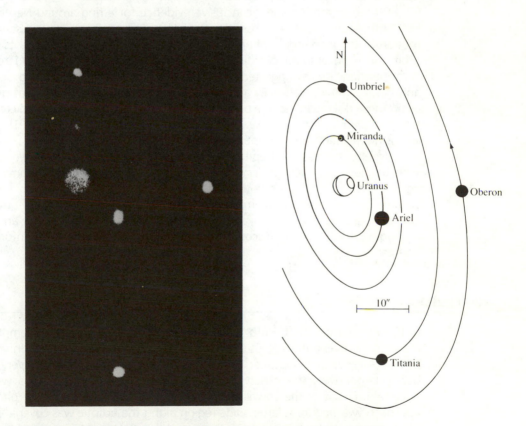

FIGURE 4.12   On the left are Uranus and its five satellites as photographed by William M. Sinton in the infrared at Mauna Kea Observatory in Hawaii; on the right is a line drawing of the planet and the satellites' orbits. (Composite photograph and drawing courtesy of William M. Sinton, Institute for Astronomy, Honolulu, Hawaii.)

wise, one would have to assign it an inclination of 98 degrees—that is, the rotation axis lies nearly in the plane of its orbit. This has interesting consequences for the planet's seasons.

Uranus was discovered in 1781 by the English astronomer William Herschel during an examination of a chart of the stars in the constellation Gemini. His first surmise that the object was a comet was proved wrong as soon as its orbit had been computed. Uranus is the first planet discovered in modern times.

An interesting discovery was made in 1977 that Uranus has nine rings revolving about it. The discovery was made because it was known beforehand that the planet was going to *occult* (pass between the earth and another object) a fairly bright star. Measurements of the duration of the occultation from more than one place on earth could lead to a more accurate determination of the planet's diameter. Some time before the occultation there were several brief drops in the combined light of Uranus and the star; after the occultation by the planet, there followed similar short extinctions. Altogether there are nine rings, all much narrower than those of Saturn. The rings have widths from 5 to 21 km (3 to 13 mi) and distances from the planet's center from 42,000 to 51,000 km (26,000 to 31,700 mi). The planet's radius is 25,400 km (15,800 mi). There is some inconclusive evidence for a ring around Neptune.

After some years of observation of Uranus's positions, it was seen that the motion of the planet did not agree with the motion determined from its computed orbit. Since no amount of adjustment of the orbital elements would remove the discrepancies, it was suggested that the motion of Uranus was being disturbed by another, unknown planet. By a theoretical analysis based on the law of gravity, and using these discrepancies, the position of the unknown planet was predicted. In 1846 the new planet, Neptune, was found close to the predicted position. The discovery was a great triumph for scientific law.

Neither Uranus nor Neptune show any clear-cut surface detail when observed with a telescope. Even under good seeing conditions and a high power the images are quite small. For both planets hydrogen and methane can be detected spectroscopically, but not ammonia, which appears to be frozen out at these low temperatures. When *Voyager 2* passes Uranus in January 1986 and Neptune in August 1989, it may discover more satellites and reveal other information of interest.

4.20 PLUTO

In 1930 the ninth planet, Pluto, was discovered. Before this date an attempt had been made to discover a trans-Neptunian planet by observing discrepancies in the motions of both Uranus and Neptune; but because these discrepancies were very small and barely larger than the observational errors, the derived predictions were highly uncertain. Finally at the Lowell Observatory at Flagstaff, Arizona, the photographic approach was begun. A large, wide region along the ecliptic was covered by pairs of photographic plates taken a week apart, and the plates were examined for a moving object with the characteristic motion of a planet beyond Neptune. In February 1930 the new planet was found and was given the name Pluto, for the god of outer darkness. The full stories of the discoveries of Uranus, Neptune, and Pluto are much too long to be discussed here, but they are recommended reading as chapters in the history of discovery.

In 1978, while high-quality photographic plates of Pluto were being examined, it was found that on some plates there was a faint image that was very close to that of the planet and which moved from one side to the other in a few days. This image turned out to be that of a satellite, which was given the name Charon. As a result it became possible to determine the mass of Pluto and that, along with Pluto's diameter, allowed determination that the planet has a density very close to that of water ice. The planet must be composed of frozen water, methane, and ammonia. The planet's surface temperature is about 40K.

Observations of the brightness changes of the sunlight reflected from Pluto show that there is a pattern repeated every 6.39 days; this must be the rotation period of the planet. The satellite has the same orbital period; hence Charon always appears to an observer on Pluto to be above the same surface point. The diameter of Pluto is about 3000 km (1800 mi), and Charon is about half as large. The distance between centers is about 18,000 km (11,000 mi). The mass of Pluto is only about 2/1000 that of our earth; that mass could not possibly have produced a detectable perturbation in the motion of either Uranus or Neptune.

The prospect of finding a planet beyond Pluto is remote but not impossible. So far the motion of Pluto has shown no measurable perturbations. One can say with considerable confidence that if a planet exists beyond Pluto it must be a very faint object; to look for it by the photographic survey method would be an enormous undertaking. Planet X might be found quite accidentally on long-exposure plates made for other reasons, but this possibility is extremely remote.

KEY TERMS

astronomical unit	greenhouse effect	escape speed
conjunction	volcanoes	oblateness
transit	asteroids	Roche's limit
elongation	Bode's law	occultation

QUESTIONS

1. When Venus is in the narrow-crescent phase close to inferior conjunction, it is possible to see the narrow crescent continued all the way around as a faint ring of light. Suggest a reason for this phenomenon.

2. Would you expect Mars to show phases? Explain.

3. Draw a diagram to show the circumstances under which the gravitational disturbances of Jupiter could convert the orbit of a comet of long period and high eccentricity into an orbit with shorter period and less eccentricity and with either direct or retrograde motion.

4. How would the seasons on Mars compare with those on the earth?

5. Describe the nature of the seasons on Uranus.

6. From the data in Table 4.1 show that when Pluto is at perihelion it is closer to the sun than is Neptune. What do you think is the probability of a collision between the two planets?

7. Why is it that Mercury appears to have no atmosphere?

8. Is there any reason to be concerned about a possible very large increase in the carbon dioxide content of our earth's atmosphere?

9. Why do the rings of Saturn disappear (as seen from the earth) when both the sun and the earth are in the plane of the rings?

5 The Second Solar Family: Comets and Meteoroids

Almost everyone has heard of—or in rarer cases seen—Halley's Comet, which was so brilliant in 1910 and which, with its long tail, is so often portrayed. There will be another chance to see it in the late fall of 1985 and the early spring of 1986. The pictures are accurate, but the emphasis on this atypical comet has given a wrong impression of comets as a whole. Most comets are very faint objects that do not reach naked-eye visibility, and their photographs are not spectacular. Hundreds of comets have been observed since 1900, but only about two dozen have reached naked-eye brightness. At any one time a half-dozen comets may be under observation.

A **meteoroid** is generally regarded as one of the relatively small but very numerous objects in orbital motion about the sun; when this object collides with the earth's atmosphere, it produces the phenomenon of a **meteor**. If the meteoroid reaches the earth's surface it is called a **meteorite**.

5.1 THE ORBITS OF COMETS

The average orbital eccentricity of the orbits of comets is much higher than that of the planets, but individual values can range from zero to unity. The shortest known period is 3.3 years and the longest may be millions of years. The orbital motion may be either direct or retrograde, and the inclinations can have values from zero to ninety degrees.

However, there is a pronounced and significant difference between short- and long-period comets if one draws the line at about 150 years. In the short-period group, the average inclination and eccentricity are much smaller than in the long-period group. Most of the short-period comets have direct orbital motion just like the planets and an average aphelion distance near the orbit of Jupiter. Among the long-period comets, the average inclination is high, the eccentricity is very large, and the orbital motions are about evenly divided between direct and retrograde.

An appreciable number of comets have orbital eccentricities slightly greater than unity (i.e., hyperbolic orbits). Does this mean that these objects make only one pass at the sun? It can be shown that all of these comets moved in highly eccentric orbits before they entered the planetary region, but close encounters with Jupiter or Saturn altered the orbits from ellipses to hyperbolas. Such comets will not be seen again. However, the reverse may take place. A close encounter with Jupiter may cause the period, eccentricity, and inclination to become less so that in time the comet acquires the characteristics of a short-period comet. A gravitational encounter with Jupiter may even cause the retrograde motion of a comet to become direct.

5.2   THE CHANGES IN THE APPEARANCE OF A COMET

As a comet approaches the sun from a great distance, it is first seen as a faint, fuzzy, nonstellar patch. This **head** grows in size and brightness, and finally it develops a short **tail**. The tail points approximately away from the sun, and it reaches its greatest length and brilliance when the comet is closest to the sun. After passing perihelion, the reverse process takes place until finally the comet fades away into invisibility. Because the tail always points away from the sun, the tail will precede the comet as the comet travels away from the sun. When the object is fairly close to the sun, the head can often be resolved into a small starlike **nucleus** surrounded by a halo called the **coma**. (See Figure 5.1.)

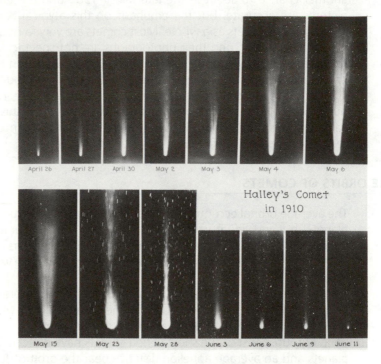

FIGURE 5.1   Fourteen photographs of Halley's Comet from April 26 to June 11, 1910. Although the comet passed perihelion on April 20, it was far from the earth and appeared relatively faint; as the distance from the earth decreased, the comet became brighter until early May and then faded rapidly as the distance from both the earth and the sun increased. (Photograph courtesy of the Mount Wilson and Las Campanas Observatories, Carnegie Institution of Washington.)

When the comet is very far from the sun, its spectrum is much like that of the sun's spectrum as reflected from a solid mass, the comet's nucleus. As the nucleus approaches the sun in the usual elliptical orbit, the fuzzy nucleus becomes larger because of the vaporization of gas and the ejection of dust particles to form the coma. This usually takes place about 3 AU from the sun. One now begins to see superimposed upon the sun's spectrum the emission spectrum of such molecules as C_2, OH, CN, CO, NH, CH, and N_2. The density of these gases is very low and some of these compounds are ionized. Most probably these gases are fragments of molecules of water (H_2O), cyanogen (C_2N_2), hydrogen cyanide (HCN), carbon dioxide (CO_2), methane (CH_4), and ammonia (NH_3). It is the above fragments that glow in their own characteristic spectrum because they have been excited by the sun's ultraviolet radiation. The fragmentation is probably caused by the disrupting action of the ultraviolet radiation and the **solar wind.** The latter is the continuous breeze, fluctuating in intensity, of protons and electrons ejected at high speeds from the sun in all directions.

The evidence is now very strong that the nucleus of the comet is a piece of ice in which dust particles are imbedded. Far from the sun the nucleus is very cold, but at about 3 AU from the sun the temperature is high enough so that the ice begins to vaporize. The dust is exposed at the surface of the nucleus so that both gas and dust are ejected to form the coma, which now grows in size as the comet approaches the sun. The pressure of the solar radiation and solar wind ejects the gas molecules and the dust from the coma to form the tail, which points away from the sun. Because the material of the tail is forever lost to the comet, the mass of the nucleus continously decreases and eventually the object ceases to be seen on its next return to the sun.

Because the nucleus contains both dust and frozen gases, one often sees a dust tail and a gas tail. Of the two forces forming the tail, that produced by the protons of the solar wind is the greater. Therefore, the impact of the protons on the light gas molecules is much greater than that of the two forces acting on the dust particles, which are more massive. Gas tails are straighter because the molecules are ejected from the coma with higher speeds than the dust particles. Gas tails usually point almost directly away from the sun, but dust tails are curved because the dust particles are moving more slowly and the speed and direction of motion of the nucleus are changing due to motion in the elliptical orbit. (See Section 1.13 for Kepler's second law.) There is now good evidence that at one end of the scale comet nuclei consist almost entirely of frozen gases with little dust, while at the other end the nucleus may be mostly dust with little ice. This could account for the widely different appearance of comet tails. The first observations of a comet from above the earth's atmosphere were made by the astronauts in Skylab in 1974 when the nucleus of comet Kohoutek passed within 80,000 km (50,000 mi) of the sun's surface at perihelion. The astronauts photographed a coma of hydrogen that was 1.6 million km (1 million mi) in diameter.

The most famous comet of modern times is Halley's Comet, which was last seen in 1910 and passed through perihelion in May of that year with a tail 160 million km (100 million mi) long. Edmund Halley, a contemporary of Newton, found that the three bright comets of 1531, 1607, and 1682 had very nearly the same orbit, which led him to believe that they were all the same object and that they returned to the sun in an elliptical orbit about every 76 years. His prediction that the object should return in 1758 was verified well after his death. After a great deal of historical research, a record has been found of every return of Halley's Comet since 240 B.C. The next return will be in late 1985 and early 1986.

In Section 4.12 it was noted that the great majority of asteroids have only slightly elliptical orbits. However, there are a few with quite elliptical orbits with a semimajor axis of only a few astronomical units and a perihelion position fairly close to the sun—some inside the earth's orbit. Because their orbits resemble those of the shorter-period comets, it has been suggested these objects were once covered by an ice layer like that in comet nuclei. This layer would have been acquired a long time ago, when it is presumed that the objects had an aphelion point far out in the solar system where Jupiter and the larger planets were formed from the gases listed in Sections 4.14 and 4.15. Such objects could return to the sun every several hundred years or tens of thousands of years. At some time these objects could have been forced into much smaller orbits with shorter periods by the influence of Jupiter and Saturn. They would look like comets when they were close to the sun, but after all the ice was gone there would be no tail; nothing but the solid core would remain, shining like a faint asteroid. It is also possible that these atypical asteroids are just ordinary asteroids whose orbits were like this from the time of their origin or that they have had their original orbits greatly altered by Jupiter.

Little can be said about the masses of comet nuclei except that they are very small. The mass of a comet can be determined only if it produces a measurable disturbance in the motion of a body of known mass. In 1886 Brooks' Comet passed through the satellite system of Jupiter without producing any observable changes in the satellite orbits. However, the pull of Jupiter changed the comet's orbital period from 29 to 7 years. One can calculate what the comet's nuclear mass must have been in order to produce observable changes in the satellite orbits, but because no changes were detected, the calculated mass must be a maximum value. From such information and the value for density of ice, one can conclude that the nuclear diameter of a comet even as large as Halley's must be no more than a few miles. No comet has ever shown a disc to an observer using a telescope. The appearance is always stellar.

5.4 METEORS, METEOROIDS, AND SWARM METEOROIDS

A meteor may appear as a very faint streak of light, or, at the other extreme, it may have the brilliance of a long fireball trail and may be visible for hundreds of miles. At night the latter may cast strong shadows. The meteoroids range in size from tiny micrometeoroids weighing but a minute fraction of an ounce to very large ones weighing many tons. The impact speed of a meteoroid with our atmosphere may be almost zero or as great as about 70 km/s (44 mi/s), depending on the relative orbital speeds and the directions of motion of the two bodies. Most meteors are first seen at an altitude of about 130 km (80 mi) and fade out at about 40 km (25 mi). Because of their tremendous speeds their kinetic energy per gram is enormous. Most of this energy is dissipated within a brief time interval ranging from a fraction of a second to two or three seconds. The energy is dissipated by pushing aside the air, exciting and ionizing molecules and atoms, and heating the surface of the meteoroid. A glowing column of air and ablated material is left behind. Hundreds of millions of meteoroids strike and burn up in the earth's atmosphere every day.

About the middle of November each year, the number of meteors shows a perceptible increase and then a decrease, the whole lasting just a few days. Most of the meteor tracks radiate from a **radiant** in the constellation Leo and are called Leonids. This divergence from a radiant is a phenomenon of perspective and is the result of the meteoroids' traveling in parallel paths in space as they strike the earth's atmosphere. During the year, it is possible to observe many swarms, each with its own radiant, although the hourly rate is very small for most of them. Note that no swarm meteoroid has ever been known to become a meteorite.

By photographic triangulation using two telescopes placed some distance apart, one can determine the motion of the meteoroid before it struck our atmosphere. From this, one can determine that all the meteoroids from the same stream follow fairly closely the same orbit around the sun. Each radiant indicates a separate swarm moving in its own distinctive orbit. Clearly, we do not become aware of a swarm unless its orbit intersects the earth's orbit at some time during the year. Because the individual meteoroids may have orbital periods of years, and because we see some each year when the earth reaches the intersection, it appears that the meteoroids are strung out in a long chain all around the orbit. On some occasions, meteors from a given swarm are observed for several nights. This means that the earth takes that long to move across the swarm diameter, which may be several million miles.

Occasionally, it is observed that a swarm meteoroid breaks up along the meteor path, and the meteor fluctuates rapidly in brightness. From this and other evidence, it appears that many swarm meteoroids are not solid lumps like grains of sand, but are more like cotton fluff. They may be made up of a mass of crystalline material. Occasionally, it is possible to obtain the spectrum of a bright meteor and to observe the emission lines of such elements as iron, magnesium, calcium, silicon, and manganese. For some of the very fast meteoroids, the energy is so high one can observe some of these metals in the ionized state.

5.5 THE RELATION BETWEEN SWARM METEOROIDS AND COMETS

The principal evidence for such a relationship lies in the numerous cases of close correspondence between a comet's orbit and a specific swarm orbit. The theory that is involved here is often called the "dirty ice theory."

According to this theory, the nucleus of a comet is made up of a chunk of ice of the several gases detected in the spectrum of a comet. Imbedded in the ice are the meteoroid particles. As the comet approaches the sun and becomes warmer, the ice sublimes to form the coma and tail, and the solid particles are exposed at the surface of the nucleus. Since the gravitational attraction of the nuclear mass is very small, only a feeble force is required to eject the meteoroid. Rotation of the nucleus and small explosions or puffs of the subliming ice would be sufficient. The particles would be ejected in all directions. Those ejected forward or backward from the direction of motion of the nucleus string out along the comet's orbit and form the chain. Those ejected sidewise broaden the chain but continue in practically the same orbit. It must be noted that the speed of ejection of the particles will always be much smaller than the speed of the nucleus. Therefore, the orbit of the meteoroids will be (after a time) practically the same as that of the comet from whose nucleus they originated. In time, planetary perturbations and solar radiation forces may alter this correspondence a great deal, and one can expect that eventually the orbits of the meteoroids will no longer resemble those of the comet.

Since the gases are lost to the nucleus, there must come a time when the ice is gone; then the only evidence that there even was a comet is the meteoroid swarm. There have been a number of cases in which a faint comet was discovered, observed for a few returns, and seen no more. It is now generally believed that at very great distances and in all directions from the sun, there are millions of cometary nuclei. Occasionally one of these is perturbed a little by a nearby star so that the nucleus begins to fall in toward the sun in a highly elliptical very long-period orbit. As it passes among the planets, there is a chance that Jupiter and Saturn may perturb the comet's orbit into one less eccentric and with a much shorter period. By this means, the inner solar system's supply of comets would be replenished.

In a related matter there are a number of known asteroids with rather highly eccentric orbits whose perihelion points are well inside the earth's orbit. The resemblance of these orbits to those of many comets suggests that perhaps these asteroids (1–2 km in diameter) were actually coated with cometary ice at one time and that they behaved like comets until all the ice sublimed, leaving only a solid core. It is not suggested that all comets have this solid core.

5.6 METEORITES AND THEIR ORIGIN

Only very rarely is a meteoroid massive enough to survive the plunge through the earth's atmosphere and reach the surface as a meteorite. The meteor phenomenon observed when this takes place can be spectacular and the flash may be seen in daylight. At times the shock wave in the atmosphere can be heard. On rare occasions, the impact will produce a large crater. Meteorites may break into many fragments on impact with the atmosphere.

Generally meteorites may be roughly classified as iron, stony-iron, and stony. The relative percentages for those seen to fall and then *recovered* are about 5, 2 and 93 percent. Some fifty minerals have been identified in meteorites and about that many chemical elements. The iron meteorites are composed mainly of iron, very small amounts of nickel, and small quantities of other elements. They have a complex crystalline structure. The stony-irons consist mainly of metallic silicates and small amounts of iron. They are very rare. The stony meteorites consist almost entirely of metallic silicates. A very rare type, in which about 5 percent of the mass consists of a dark, carbon-rich, tarry substance, is called a carbonaceous meteorite. This type is very fragile and most probably fragments into small pieces after entering the earth's atmosphere. They may be very common as meteoroids. Most of the tar consists of amino acids of types found on earth, but the evidence is now clear that these compounds are not the result of contamination after reaching the surface and that they were not formed by life processes. It is also clear that these carbonaceous meteorites have never been very hot. The presence of the organic compounds is a mystery.

The phenomenon of observational selection is well illustrated by the numbers of the first and third classes found in museums. The great majority of all accidental meteorite finds are of the iron variety, but most of those *seen to fall and then recovered*, are of the stony type. The main reason for this is the low breaking strength of the stony type; as a result they shatter into small pieces, while the irons, with their high tensile strength, tend to survive the experience as one piece. Micrometeorites are defined as those so small that they do not burn up, but are slowed by numerous collisions with air molecules and then settle slowly to the surface.

The age of meteorites may be obtained by the uranium-lead method. The oldest meteorites have an age of about 4.5 billion years. It should be understood that the age refers to the length of time since the body assumed its present crystalline solid form. Note that the 4.5-billion-year age is very close to that usually given for the age of the earth and for the oldest moon rock samples. Some important event must have taken place about 4.5 billion years ago.

No swarm meteoroids have ever been known to reach the earth's surface. Their masses are too small to survive to the surface. While many shower meteoroids have retrograde orbits, there is no known case of a meteorite with a retrograde orbit. Meteorite orbits usually have small inclinations and eccentricities, and therefore they have planetary characteristics. The present theory of the origin of meteorites is that they were formed either by the disintegration of a small planet or by the collision (on at least one occasion) of two asteroids. The latter event would produce a great shower of direct orbital motion fragments, some of which would have orbits with eccentricities that would move them inside the earth's orbit at times, causing collisions with the earth. Also, the minerals in most meteorites are mainly of the type not formed under great pressures of the magnitude that would be found deep in a large body like the earth. These lower pressures would be typical of the interiors of the larger asteroids.

5.7 METEORITE CRATERS

All over the world there is evidence of craters produced by meteorite impact. Craters range in diameter from a few feet to a few miles. There may be a fossil crater in South Africa with a diameter of 64 km (40 mi). It could have been produced by a small asteroid. Almost all meteorite craters are nearly circular in outline. The shape of the crater seems to have little to do with the angle of impact. The shock wave of highly compressed air on the forward side of the meteorite plays a major role in blasting out and creating a circular crater much larger than the impacting body. The craters are much like those produced by an aerial bomb. However, not all circular craters are meteoritic in origin; several geologic processes can produce the same effect. One kind of evidence is in the presence around the crater of meteorite fragments that were blasted out in the explosion. A good example of this is Arizona's Barringer crater, which is 1200 m (4000 ft) in diameter. (See Figure 5.2.) The rock formations around the edge of this crater have been badly fractured and turned upward by the force of the underground explosion. One of the best indicators is the presence of the mineral **coesite**, which is formed from quartz under very high pressures. Its presence in a crater is now thought to be proof positive of the crater's meteoric origin. Some of

FIGURE 5.2 The Barringer crater near Winslow, Arizona. The crater is 4000 feet across and about 600 feet deep. The interior walls are very steep. The far wall of the crater shows lines marking the sedimentary beds of sandstone in which the crater was formed. The age of the crater is unknown, but it is estimated to have been formed about 50,000 years ago. (Photograph courtesy of Yerkes Observatory, University of Chicago.)

the large circular lakes in eastern Canada are now confirmed as meteorite craters, in part because of the presence of coesite.

5.8 THE METEOROID HAZARD TO SPACE TRAVEL

Any space vehicle beyond the earth's atmosphere will be bombarded by many meteoroids of all sizes, but the hazard has been much exaggerated. The outer shell of the vehicle would stop all micrometeoroids. A collision with an object the size of a pea would be damaging, depending upon where it struck the vehicle. A collision with one weighing several pounds would prove fatal to the mission, no matter where it struck. However, the space density of meteoroids large enough to seriously affect the vehicle and its occupants is so low that no serious problem is presented. The hazard per mile to space travel of a seriously damaging collision with a meteoroid is so much less than the hazard per mile to the operation of an automobile that even a trip to and from the moon may be undertaken in comparative safety.

5.9 THE 1908 SIBERIAN METEORITE(?)

On June 30, 1908, an extremely brilliant daylight meteor was observed in the Tunguska River area in central Siberia. The atmospheric shock wave rattled windows as far as 480 km (300 mi) from the impact point and in time was registered on microbarographs all around the earth. Small tremors were felt up to 160 km (100 mi) from the impact area. In the middle 1920s, a Soviet government expedition to the area found craters up to 45 m (150 ft) across. As far as 32 km (20 mi) from this area, trees were found to have fallen with their tops pointing away from the crater area. Oddly, no trace of meteoritic material has ever been found, even though it seems perfectly clear that most of the effects observed were caused by a body from outer space. There is now some agreement that the object was not the usual large meteoroid, but the nucleus of a small comet. The heat produced on impact with the atmosphere and the earth's surface vaporized the ice of the nucleus and the small dust particles imbedded in the ice, so nothing solid remained. It is doubtful if the object was larger than 30 m (100 ft) in diameter.

A recent study of this phenomenon suggests the collision with the earth's atmosphere of an asteroidal body with a diameter in the range of 90 to 190 m (300–600 ft). This extremely violent collision fragmented the body into a vast number of minute particles, most of which were blown into the earth's atmosphere and scattered in due time over most of the earth's surface. Ice cores from the Antarctic cap shows a pronounced increase in the number of micrometeoroids within a short time after 1908. Although there is some disagreement about the nature of the impacting body, it is clear that it originated as an extraterrestrial object.

5.10   THE ORIGIN AND EVOLUTION OF THE SOLAR SYSTEM

With this background of details about the solar system, the problem arises as to its origin and evolutionary changes. In Chapter 6 the formation and evolution of stars will be treated. Because a star is a single body, its origin and evolution present a much simpler problem than does the history of the solar system. Several hypotheses about the origin and evolution of the solar system have been proposed, but the only one thought to be reasonably successful is the protoplanet hypothesis discussed below.

This hypothesis reasonably explains the common direction of revolution of the planets, their small orbital inclinations, and the changes in density and mass with distance from the sun. With two exceptions (Venus and Uranus), the rotation direction is the same as the direction of revolution about the sun. These may be just accidents, but if the protoplanet hypothesis is close to the truth, there ought to be many other planetary systems around other stars. Because there is so much detailed information about our solar system, we cannot expect that any hypothesis will explain every detail. Some of the more important unexplained details will be discussed at the end of this section.

The proto-planet hypothesis assumes the existence of a cold, slowly rotating interstellar globule of gas and dust very much larger than the present limits of the planets. This globule began to contract under its own gravitational attraction and began to rotate more and more rapidly until it became a flattened disc with a massive central condensation which finally formed the sun. This increased rotation rate is analogous to the action of a spinning ice skater.

This theory is also sometimes called the **planetesimal hypothesis**. This is because it is assumed that the dust particles somehow accreted to form small bodies (the planetesimals), which by further collisions formed even larger ones. The actual process is not fully understood. In this turbulent rotating mixture of gas, dust, and particles, some of these planetesimals by chance grew to considerable size at the expense of the smaller ones and acquired atmospheres from the interplanetary gas. These objects are called proto-planets, and there must have been many more of them than the present number of planets. As time went on, these proto-planets in their revolution about the sun swept up planetesimals of all sizes. At present, only a small fraction of the latter remain; the asteroids may possibly be some of the smaller proto-planets.

While the proto-planets were becoming larger, most of the globular mass was contracting at the center to eventually form the sun. As this mass contracted, the loss of gravitational energy was converted to heat, and the proto-sun's central temperature increased enough to begin the thermonuclear conversion of hydrogen to helium and the consequent release of mass energy. At some point, the heat generated in this way produced throughout the sun a gas pressure sufficient to balance the gravitational attraction; it ceased to contract and became a star (our sun). This theory of star formation predicts that in its last stages, the star is unstable and ejects some of its mass into space at high speeds. The force of this gas and the sun's radiation pressure would eventually sweep clear from the solar system that gas which had not been captured by the proto-planets. The interplanetary space would become transparent and the sun's heat would warm the planetary atmospheres by an amount decreasing with distance from the sun. It is thought that for the planets Mercury through Mars (including the moon), with their low escape speeds, the original atmosphere (mostly hydrogen) escaped. The present atmospheres of Venus, Earth, and Mars are thought to have been formed by later volcanic processes. Because the temperatures of the planets Jupiter and beyond are very cold, they retained even hydrogen and helium to attain their present large masses and low average densities.

The origin of the satellites is rather obscure and this is particularly true for our moon. It is not generally believed that our moon was formed by fission from a rapidly rotating earth or that the earth-moon system was a double proto-planet. More likely, the moon was formed as a separate planet and captured by the earth in some process not well understood. Some of the satellites of the more massive planets, like those of Jupiter and Saturn, may have been small proto-planets formed at the same time and place as the larger body, or they may have been captured later. The correct answer is uncertain.

Earlier it was suggested that the shorter-period comets originate from a large swarm of cometary nuclei in the far outer reaches of the solar system beyond the planets. Most of these nuclei appear to consist of ice condensations of several gases mixed with dust in a wide range of proportions. It is possible that when the solar nebula was quite large, the molecules were slowly forming out of the gaseous atoms at temperatures much lower than those found in most chemical laboratories.

Another aspect of this hypothesis to be discussed here is the formation of the asteroid belt. Possibly, the larger of the present asteroids are some of the smaller proto-planets never swept up by the larger planets like Jupiter. With their small surface gravity and low escape velocity, they lost whatever atmospheres they had when the interplanetary space became transparent to the sun's heat. However, the existence of so many small asteroids in the range from 2.5 to 3.3 AU needs an explanation. Some of these may have been planetesimals that were prevented from forming a larger planet because of the disrupting action of Jupiter. Or they may have been formed (on at least one occasion) by the collision of two of the larger asteroids; the fragments spread out somewhat and continued to revolve in the same direction around the sun.

With respect to meteoroids it should be noted that those we see as meteorites are obviously fragments of a larger mass and also that *all* meteorites show a crystalline structure. The first observation suggests a collision between two asteroids to form smaller fragments. The second demands that the crystals were formed from a cooling melt of the minerals. A third observation is that there is a wide variety in the chemical composition of the meteorites, as though they came from a chemically differentiated body. See Section 2.7. The second and third observations require that the colliding bodies were fairly large and that they were probably two large planetoids (asteroids?) a few hundred miles in diameter. As these bodies formed from smaller ones they became heated by impacts, chemical differentiation took place slowly, and the bodies cooled and eventually collided to fragment into smaller meteoroids. Another point is that the minerals found in meteorites are of a kind not formed under the high pressure that would exist for a large planet like Mars or the earth.

5.11 SOME UNSOLVED PROBLEMS

One of the most important unsolved problems is the mechanism by which the dust particles would adhere after a collision to form larger particles.

The inner four planets and our moon were probably hot enough in their early history for the heavier minerals to sink to the planet's center, and the lighter ones concentrated in the surface layers. Geologists call this "differentiation." But how did the planets become hot? Probably this took place as the proto-planet cores were being bombarded by the planetesimals and their energy of motion (kinetic energy) was converted to heat. This must have taken place in a relatively short time span so that the heat was generated by collisions more rapidly than it radiated away. As the supply of planetesimals was depleted, a crust formed which became thicker with time. The heat produced by the radioactive disintegration of uranium and thorium made some contribution to this heat, but it is doubtful that this source was a major contributor.

It is peculiar that the oldest rocks on the moon have an age of about 4.5 billion years and that they originated from the crust that formed as the moon cooled. However, there is evidence that most of the craters were formed by impact in a fairly short time span beginning about 4 billion years ago. What is the origin of these large meteoroids, and why did the formation period wait for about a half-billion years after the lunar crust was formed? Could these meteoroids be smaller asteroids produced by the collision of two larger asteroids as suggested earlier? Another lunar problem is that the seas appear to have been formed in about a half-billion-year interval centered on 3.5 billion years ago. What caused the sea formation to have been the last major lunar event?

One of the most important problems is why the sun rotates so slowly—a period of about 27 days. If the sun formed as described earlier, its rotation period should be only a few hours. One proposal is that at one time the sun had a strong magnetic field which interacted with the ionized gas of the solar nebula and produced a braking action on the sun's rotation. There is no clear-cut answer to this problem.

It should be reasonably clear that the proto-planet hypothesis is a fairly satisfactory one, but a number of problems remain to be solved. If this hypothesis is reasonably close to the truth, then there ought to be many other stars with planetary systems of their own and some of these planets may be inhabited by living forms.

KEY TERMS

meteorite	head	solar wind
meteoroid	tail	radiant
meteor	nucleus	planetesimal hypothesis

QUESTIONS

1. What effects would you expect if the earth passed through the tail of Halley's Comet?

2. Attempts have been made to observe the impact of a meteoroid on the moon by looking for the flash on the dark half of the moon at first and third quarters. No impact flash has ever been seen. Suggest a reason for this failure.

3. Why do some meteor showers last as long as ten days while others persist for only a day or less?

4. What is the difference between the dust tail and the gas tail of a comet?

6 The Sun: An Introduction to the Stars

Our sun is the nearest star, and it is the only one which is seen as a disc instead of an unresolved point of light. It is possible to examine the sun's disc in great detail—this can be done for no other star. The sun is almost the sole source of radiant energy in the solar system, and of course life on earth would be impossible without it. The light from the stars is negligible compared with that from the sun. The presence of fossils in very old rocks is evidence that the sun's surface temperature and the energy we receive from it have changed but little in the last half billion or so years. For most of its life the sun has obtained its energy from the conversion of hydrogen to helium.

6.1 THE PHYSICAL STRUCTURE OF THE SUN

The mass of the sun is 2.2×10^{27} tons, 330,000 times more massive than the earth. The sun contains 99.8 percent of the mass of the solar system. Its diameter is 1,392,000 km (864,000 mi), and its average distance from the earth is not quite 150 million km (93 million mi, 1 AU). The mean density of the solar material is 1.4 times that of water. Because the sun is so hot throughout all its volume, all of its matter must be in the gaseous state.

The visible surface of the sun is called the **photosphere** (Greek for "light sphere"). Above it, the gases are almost entirely transparent, and below it they are opaque. Just above the photosphere is a layer called the **chromosphere** (Greek for "color sphere") because of its reddish color, which can be seen when the photosphere is hidden during a total solar eclipse. The lower level of the chromosphere is often called the **reversing layer** because it is the main source of the absorption lines in the solar spectrum. In the chromosphere and extending through it well up into the corona are the **prominences** (red in color), which can be seen at the sun's edge (the limb) during a total solar eclipse. At the same time, one can observe the **corona**, a faint white solar halo, which has been observed at times to be several solar radii above the photosphere. None of these layers is sharply bounded, but instead they merge into one another. The average temperature of the photosphere is 5750K

(10,000°F). At this temperature, the maximum of the spectral energy distribution is in the visible region, and here is found a large fraction of the total solar radiation.

Theoretical studies have shown that the sun's interior temperature rises rapidly below the photosphere and probably reaches 15 million K at the center. Not far below the photosphere, the temperature is so high that most of the atoms are completely ionized. As we will discuss in the next chapter, it now seems obvious that the sun's energy source is the nuclear transformation of hydrogen into helium. Since the mass of the sun is great and the nuclear process is a very efficient one, it appears that the sun is capable of shining with its present brilliance for many billions of years.

6.2 THE SURFACE PHENOMENA OF THE SUN

Early visual observations and the long record of later photographic data have revealed an incredible amount of detail about the sun. It is regrettable that such diverse information can be obtained for no other star.

Telescopes carried by balloon to 24,000 m (80,000 ft) or more, well above most of the disturbances of the earth's atmosphere, reveal in detail the photospheric **granules** with diameters of about 1600 km (1000 mi). They have a honeycomb appearance of bright spots surrounded by darker boundaries. A particular granule has a lifetime of only a few minutes. From Doppler shift observations it is clear that the bright central region of a granule is a rising column of hot gas which cools by radiation, the darker gas then sinking back down around the edge of the granule to become heated again and rise once more. These granules are an important mechanism for the transport of heat from the solar interior.

The most obvious features are the **sunspots.** See Figure 6.1. The smallest spots are about 1600 km (1000 mi) across; the largest have diameters of 160,000 km (100,000 mi) but are rare. The larger ones can be seen by the naked eye through fog, smoke, or a dark absorbing glass. Sunspots are roughly circular structures consisting of two well-defined parts. The central, darker region is the **umbra** and the outer, brighter ring is the **penumbra**. The umbra is about 2000K cooler than the photosphere outside the penumbra. Because of its much lower temperature, the umbra radiates a smaller luminous flux than the photosphere and is darker in contrast. If the umbra could be seen by itself, it would be intensely bright.

Most spots occur in pairs or in groups dominated by two large spots. The line joining the centers of a pair is about parallel to the sun's equator. From observations of spots, it is known that the sun does not rotate as a solid. The rotation period at the equator is 25 days and increases to 27½ days at latitude 30 degrees. Since spots are rarely seen beyond this latitude, the rotation period is found from Doppler shift observations of opposite limbs of the sun. The period increases to 35 days at latitude 75 degrees. Almost all spots are found between latitudes 5 and 35 degrees in each hemisphere. They are scarce in the narrow equatorial region and in the polar zones.

The lifetimes of spots range from as short as a day to a few months for those that grow to large size. A spot is first observed as a dark dot about 1600 km (1000 mi) across on the photosphere. Most spots fade away in a day or two, but those that persist may grow into groups 320,000 km (200,000 mi) across. One of the most interesting aspects of sunspots is their magnetic polarity. Field strengths comparable to that of a high-quality permanent magnet are observed.

THE SUN: AN INTRODUCTION TO THE STARS

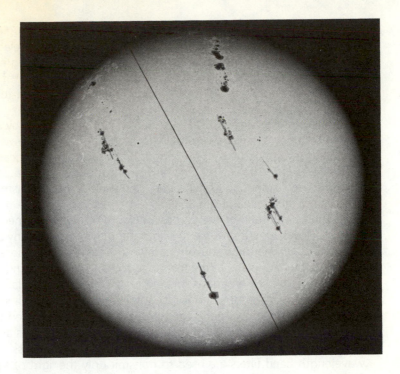

FIGURE 6.1 A photograph of the sun taken at sun spot maximum in December 1957. Observe the individual spots and the spot groups. The straight line is the approximate position of the sun's equator. Note how the spots are strung out across the sun's surface in two bands on either side of the equator and approximately parallel to it. (Photograph courtesy of the Mount Wilson and Las Campanas Observatories, Carnegie Institution of Washington.)

6.3 PROMINENCES

Under the conditions of a natural or artificial total eclipse, it is often possible to see large bright streamers or clouds above the photosphere, with their bases in the chromosphere and near the limb of the sun. The clouds are composed largely of hydrogen, but they contain some calcium and other elements. The hydrogen clouds are red because so much of the emission is at one of the characteristic emission lines of hydrogen, at the wavelength 6563 Å in the red. The **quiescent** type of **prominence** may cover the sun's surface for tens of thousands of miles and be equally as high. Sometimes it persists for days without much change in appearance. The **eruptive** variety may be "blown" off the sun as fast as 650 km/s (400 mi/s) and finally disappear a million miles above the photosphere. The **fountain** prominences show inward and/or outward motions from a spot group. At one time, the prominences and the corona could be studied only at the time of a total eclipse of the sun and only for the few minutes of the eclipse duration. Astronomers now have instruments for the production of artificial eclipses, so the prominences and the corona can be studied on any clear day. A network of these solar observatories around the world often makes it possible to obtain a 24-hour record of the sun's behavior. The reason for prominences is not clearly known. For some years, solar astronomers have been working on the problem of the correlation and interdependence of all solar phenomena. One of their most fascinating exhibitions is a time-lapse motion picture of a prominence, covering the motion of a prominence for a period of an hour or more with a projection time of only a few minutes. See Figure 6.2.

The prominence seen in Figure 6.2 was photographed by a *spectroheliograph*. This instrument is capable of obtaining a photograph of the whole sun or a prominence at just one particular wavelength. In this case the light came from the calcium spectral line at 3933 Å and shows only the distribution of calcium in the prominence. To study the distribution of hydrogen, one usually sets the instrument for the bright hydrogen H-alpha line at 6563 Å.

The photograph of the corona in Figure 6.3 was obtained at a total eclipse of the sun. However, it is possible to photograph the corona in any clear sky by means of a *coronagraph*. Here the sunlight coming from the photosphere is blocked by a disc placed in the focal plane of the telescope so that the light of the corona and the edge prominences passes on and is imaged on a photographic plate or film. This instrument is used because the surface brightness of the photosphere is at least a million times that of the corona, and the photospheric light scattered by the earth's atmosphere prevents seeing that of the corona. These instruments are usually placed at a high altitude where the atmospheric scattering is much less. In addition, narrow-wavelength band filters are used to transmit only the light of selected spectral lines and to reject all other wavelengths.

Solar observatories orbiting the earth well above the atmosphere avoid both the

FIGURE 6.2 A photograph of the sun taken with an artificial eclipse. Observe the bright ring of the chromosphere and the several kinds of prominences distributed around the sun's limb. (Photograph courtesy of the Mount Wilson and Las Campanas Observatories, Carnegie Institution of Washington.)

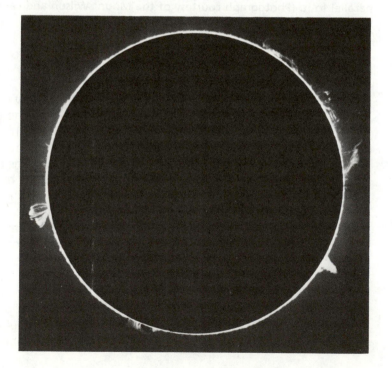

FIGURE 6.3 The solar corona photographed at the total solar eclipse of June 8, 1918. Observe the streamers in the corona and a few small prominences at the moon's edge. (Photograph courtesy of Mount Wilson and Las Campanas Observatories, Carnegie Institution of Washington.)

light scattered by the air and the absorption, so now any spectral region can be studied. The Skylab astronauts were able to obtain a wide variety of photographs of many solar phenomena.

6.4 SOLAR FLARES

One of the most spectacular and short lived of all solar phenomena is the *solar flare*. Flares appear as very bright spots in the photosphere, and they are usually near a spot group. A typical flare reaches maximum brightness in a few minutes and then fades away more slowly, disappearing in about a half hour. Flare temperatures are much higher than those of the photosphere, and they emit large amounts of ultraviolet light. The reason for flares is not well known, and it is not possible to predict their occurrence. However, they are known to be more common when sunspots are more numerous. The intensity of the **solar wind** often increases very sharply at the time of a solar flare and may be as much as a hundred or more times intense. Since the solar wind particles do not travel with the speed of light, it may take a day or two after the flare outburst before the effect is observed on the earth.

The sun's percentage change in brightness during a flare is very small, but if the same flare were to occur on a star with one-tenth the diameter and hence one-one hundredth the area of the sun, the change in brightness would be easily perceptible. In recent years, discoveries have been made of a number of "flare stars" in which there may be a sudden increase in brightness (in a matter of minutes or seconds) and then a slow fading. These stars are normally red and are dwarfs with respect to the sun's size. During a flare, the color changes from red to blue, and the star may become a hundred times brighter. It is probable that this phenomenon is a local flare much like the solar flares, except that on these dwarf stars, the flare has a surface area not much smaller than the star itself. There are some data to show that at these times there is also a radio noise flare, just as there is when a solar flare takes place.

6.5 THE SOLAR CORONA

The corona is a large white halo around the sun and extends above the photosphere for several million miles. Its total light is about one-millionth that of the photosphere. The **outer** corona exhibits a spectrum like that of the sun, indicating that it is made up of small reflecting particles.

The **inner** corona is more interesting. Its spectrum consists of rather broad emission lines superimposed on a continuous spectrum with no absorption lines. For many years the wavelengths of these emission lines could not be identified with those of any known chemical elements, and for a time there was serious talk of a new element named "coronium." It was later discovered that these lines come from very highly ionized atoms of such elements as iron, calcium, and nickel. Some of these atoms have lost as many as 15 electrons. One way to produce atoms which have lost so many electrons is for the coronal temperature to be so high that the thermal velocities of the atoms would cause intense collisional ionization. This would also account for the broad emission lines; their normal width would be greatly increased by Doppler broadening caused by the very large radial velocities of the emitting atoms in the coronal gas. If equipartition of energy existed between the relatively massive ionized atoms and the much less massive but relatively numerous electrons, the average electron velocity ought to be far higher. The scattering of the sun's photospheric light by these electrons would result in such a large Doppler broadening that the solar absorption lines would be almost completely smeared out and the scattered solar spectrum would look like a continuous spectrum. This type of reasoning now leads astronomers to believe that the temperature of the inner corona is about 1 million K, far higher than that of the photosphere. It is not at all clear just how this elevated temperature is maintained. In solar flares the spectral lines of iron, from whose atoms 25 electrons have been removed, have been observed.

It is now known that many stars, particularly those of spectral types G and K (see Section 6.16), also have extensive coronas and for some the corona is much larger than that of our sun. Some stellar coronas have been detected to be much stronger X-ray sources than our sun.

6.6 THE SUNSPOT CYCLE

It has been known since 1851 that the number of sunspots has a rough periodicity, with an average of nearly 11 years. At spot minimum, there may be several successive days when no spots are seen. The plot of sunspot numbers against time shows a relatively rapid rise to maximum from minimum in a time of 4 to 5 years and then a slower decline to the next minimum. At sunspot maximum, it may be possible to see 100 spots at a time. The general rapid rise to maximum and slower decline to minimum is characteristic of all cycles, but the time between successive maxima varies from 9 to 14 years. (See Figure 6.4.) Sunspots are almost always found in the range from 5 to 35 degrees of latitude in each hemisphere. Hence, they avoid a narrow zone at the equator and the polar regions.

If we consider only spot pairs, we find that the line joining their centers is about parallel to the sun's equator, and so we may speak of a "lead" and a "following" spot in a pair. In any one cycle, the magnetic polarity (N or S) of the lead spot will

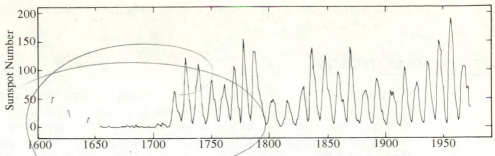

FIGURE 6.4 Annual average sunspot numbers from the data which Galileo began to gather after 1610 through the concluding data for about 1974. Note that the sunspot numbers obtained by Galileo and others after him decreased until about 1645 and then remained near zero until about 1715, after which the presently familiar pattern began. The period of about 70 years from 1645 to 1715 is called the Maunder Minimum. (From June 1976 issue of *Sky and Telescope* as adapted from John A. Eddy, "The Maunder Minimum," *Science* 192 (1976): 1189–1202. Copyright 1976 by the American Association for the Advancement of Science.)

be the same for all pairs in one hemisphere and the reverse in the other. In the next cycle, the polarity of the lead spots will reverse between hemispheres from what it was in the preceding cycle.

At the beginning of a new cycle around spot minimum, the new spots of that cycle will appear in small numbers around latitude 30 degrees in both hemispheres. As the cycle progresses and the number of spots increases, older ones die out and newer ones appear at progressively lower latitudes, with the result that during a cycle the average spot latitude decreases to about 5 degrees at the cycle's end. Toward the end of a cycle, there is an overlap of about 2 years between cycles; during this time one can see spots of the old cycle around 5 degrees latitude and spots of the new cycle around 30 degrees latitude.

Astronomers have become so accustomed to the present 11-year sunspot cycle that they have tended to ignore the evidence that this cyclic behavior has not always existed. Starting in 1610, Galileo first saw spots with a telescope and began to make occasional counts of their number each day. Not until 1851 did Heinrich Schwabe pull all the available data together and demonstrate the 11-year cycle. About 1890 Gustav Spörer and E. W. Maunder pointed out that the early record of the observed number of sunspots showed that there were scarcely any spots in the interval from about 1645 to 1715 and that the 11-year cycle was absent. This period is now referred to as the Maunder Minimum. This minimum has now been amply confirmed not only by observation of an extreme paucity of spots in this interval but also by a conspicuous absence of the *aurorae borealis* (northern lights) and the constancy of the width of tree rings during that period. It is also during this period that Europe experienced its Little Ice Age, when mean annual temperatures were conspicuously below normal, but this may have been the result of other causes. Coincidentally, many volcanoes were erupting in this period and the dust may have partially obscured the sun. Recent studies show that when Galileo first began observations of sunspots the numbers were about average, but on the downswing, and that there had been another minimum in the approximate time interval from 1460 to 1550 A.D. Thus, it would appear that the sun is not constant in all its aspects, but there is no explanation for the above behavior.

6.7   THE SOLAR SPECTRUM

The spectrum of the solar disc is of the absorption-line type. Because the sun is so bright, it is easy to obtain its spectrum at very high dispersion, which reveals a great deal of detail. Thousands of absorption lines have been measured. When their wavelengths are compared with laboratory values, about 60 chemical elements reveal themselves in the reversing layer. There may be some elements whose lines show only in the ultraviolet or infrared parts of the spectrum which are cut off by the earth's atmosphere. Rocket and satellite observations from above the atmosphere will probably settle this question. Most of the absorption lines result from neutral atoms, some are caused by atoms easily ionized at the sun's temperature, and others are produced by molecules, particularly in the cool region of sunspots.

6.8   RADIO NOISE FROM THE SUN

During World War II, radar instruments showed that the sun emits radio noise. It is now known to do so over a wide range of wavelengths. The opacity of the solar atmosphere for radio waves increases with increasing wavelength. At values of 10 to 20 meters, the apparent size of the sun is large because the energy comes from the extended corona. For wavelengths around a few centimeters (an inch or two), the corona is transparent, and the sun appears smaller because these shorter waves come from just above the photosphere.

Around spot minimum, the radio sun is "quiet." The radio brightness is about that to be expected from a body of the sun's photospheric temperature. At spot maximum, the noise level is much higher and subject to very large fluctuations. At such times, enormous bursts may occur with an intensity a million times that of the quiet sun.

6.9   THE EFFECTS OF SOLAR ACTIVITY ON THE EARTH

It has already been mentioned that at the times of maxima in the sunspot cycle and after solar flares the intensity of the solar wind tends to increase, often by large amounts. One result of this increased activity is a considerable interference with radio communications. The great increase in the ultraviolet radiation causes a large increase in the ionization of the upper atmospheric gases. This produces disturbances in the radio reflectivity of the ionospheric layers because radio waves are reflected by an ionized gas. Because the particles in the solar wind are charged particles and moving charged particles are an electric current, bursts in the solar wind give rise to magnetic storms. These storms can easily be detected by the oscillations of a delicately mounted magnetic compass needle. Another interesting effect is that the width of the annual rings of trees growing in semiarid climates is greater near sunspot maximum than it is around the minimum. The effect is real, but the reason is unknown. One might expect that the width of the tree rings would depend on the rainfall, but so far there appears to be no direct correlation between the sunspot number and the weather.

In addition to a permanent feeble glow in the earth's atmosphere called the permanent aurora, at the times of solar flares or unusually large sunspots one can observe brilliant displays of the aurora borealis. At the same time in the southern hemisphere observers can see the *aurora australis* (the southern lights). These brilliant displays are due to the bombardment of our atmospheric gases by ultraviolet photons and high-energy protons and electrons emitted by solar flares. At these times the increase in the intensity of the solar wind may be a thousand times.

6.10   AN INTRODUCTION TO THE STARS

Before we begin our study of stars as individuals or groups, it is desirable to understand a number of concepts and techniques and to learn a certain nomenclature. Our sun is an average sort of star in many ways. There are stars as much as 100,000 times brighter or fainter than our sun. Some dwarf stars are only 1/100 the solar diameter, and a very few are so large that they could include the whole of the earth's orbit. The range in mass seems to be from a little less than 1/100 the sun's mass to about 60 times as great. Surface temperatures range from as low as 2500K to as high as 100,000K.

6.11   CONSTELLATIONS

Modern astronomers have inherited from the dim past the custom of dividing up the whole sky into 88 areas called **constellations.** Most of the constellation names and forms have come to us from the Greeks, preserved for us through the Dark Ages by the Arabic civilization. Most of the names are Latin translations of the Greek names. These groupings are useful, just as it is useful to refer to a general area of the United States such as the Pacific Northwest or the Gulf Coast region. If you tell an astronomer that you are "working in Cygnus," he will know that this is a constellation about halfway between the north celestial pole and the equator in the northern hemisphere, that it is best seen in the summer sky, and that it is also a constellation in the Milky Way.

Most of the star configurations in the constellations do not look like the objects after which they were named—they were never intended to do so, but were only named in honor of those objects. Some constellations, such as Corona Borealis (the northern crown) and Lyra (the harp), do resemble the objects.

The naming of stars is such a complicated subject that reference will be made to only two systems. Many of the brighter stars have proper names such as Vega, Arcturus, Alferatz, or Betelgeuse, and often they are Arabic names. Another system indicates the star's location by assigning it a Greek letter and following this by the genitive of the constellation name. The brightest star in a constellation is usually called alpha, the next beta, and so on. As an example, Vega is the brightest star in the constellation Lyra. Its "Greek letter designation" would be alpha Lyrae.

The study of constellations and star names can be a rewarding pastime for the nonprofessional. Although a useful set of star charts appears in Appendix 3, those interested in charts of greater detail should purchase a good star atlas. Some are listed in Appendix 1.

6.12   STELLAR DISTANCES—THE STELLAR PARALLAX

As seen in Section 1.10, the parallax of a star in arc-seconds is the reciprocal of its distance in parsecs. In general, except where it would be awkward to do otherwise, we will give distances in light years. One parsec is 3.26 light years. Stellar distances are of the greatest importance to astronomers because they enable them to understand the important spatial arrangement of stars and to determine certain of their physical properties. The direct method for the determination of stellar parallax consists in obtaining over a period of a few years a series of photographs of the star and its comparison field. These photographs are taken about six months apart at opposite ends of that diameter of the earth's orbit perpendicular to the direction of the star. The change in position of the parallax star on the plate is measured with reference to a group of fainter comparison stars assumed to be more distant. By this technique, the smallest reliable parallax that can be measured is about 0".020, which corresponds to a distance of 50 parsecs, or about 160 light years.

6.13   RADIAL VELOCITY AND PROPER MOTION

The motion of a star in our line of sight, known as its radial velocity, is obtained by measuring the wavelength shift of the lines in the star's spectrum on a photographic plate with respect to some comparison spectrum, such as that of iron. Individual line shifts are converted into line velocities and the average is taken. This value is then corrected for the motion of the earth in its orbit, giving the star's radial velocity with respect to the sun. (See Section 3.14 on the Doppler effect.)

A star's **proper motion** is its yearly change in direction, measured in seconds of arc, and results from the sun's space motion, the star's space motion, and the direction and distance to the star. On the average, the proper motion of a star will increase with decreasing distance from the sun; thus, a good statistical way to look for nearby stars is to find those with large proper motion.

6.14   THE STELLAR MAGNITUDE SYSTEM

In the second century B.C., the Greek astronomer Hipparchus made a star catalogue in which he assigned numbers for brightness—1 for the brightest stars and 6 for the faintest. Much later, it was shown that the light from an average first magnitude star is close to being 100 times greater than that from the average sixth magnitude star. It is now known that the eye's response to light is not linear, but logarithmic. About 1850, astronomers made the arbitrary decision that a difference of brightness of five magnitudes was to correspond *exactly* to a luminosity ratio of 100/1. As a result, if two stars differ in brightness by *one* magnitude, the ratio of their luminosities (I_1/I_2) is the fifth root of 100, which is only slightly less than 2.512. Therefore, for a magnitude difference Δm (read *delta m*), the ratio of their luminosities is given by 2.512 raised to the power Δm. This can be expressed as follows:

$$I_1/I_2 = (2.512)^{\Delta m}$$

If $\Delta m = 1$, the luminosity ratio is 2.512, and if the magnitude difference is 2, the luminosity ratio is 2.512×2.512, or about 6.25. The following table gives the approximate luminosity ratio for several values of Δm.

Δm	I_1/I_2
0.0	1.00
1.0	2.50
2.0	6.25
3.0	16.00
4.0	40.00
5.0	100.00
6.0	250.00

The actual magnitude number assigned to any object increases as the object becomes fainter. This may seem like an odd way of expressing stellar brightness, but it is convenient because it expresses large luminosity ratios as relatively small numbers. For example, consider a luminosity ratio of 100 million (100,000,000) between two stars. This figure can be written as $100 \times 100 \times 100 \times 100$. Each 100 corresponds to a magnitude difference of 5, and there are four 100s. Hence, the luminosity ratio of 100 million is the *sum* of four 5s, or 20 magnitudes (a much smaller number).

A star's magnitude depends on its temperature or color. Hot stars are brighter in the blue than in the yellow-green (visual) region, and red stars are even fainter. Thus one can speak of a star's ultraviolet, blue, visual, or red magnitude. We shall confine our use of magnitudes to the visual.

6.15 APPARENT AND ABSOLUTE MAGNITUDES—THE DISTANCE MODULUS

Apparent magnitudes are the measure of stellar brightness as we see it in the sky. Because of the great distances between stars, apparent magnitudes are not usually a measure of the true (intrinsic) brightness of one star relative to another. If a star could be brought closer to us, it would become brighter and its apparent magnitude (m) would have a smaller number. Because stars have a great range in intrinsic luminosities, it is desirable to have a system whereby this luminosity can be expressed in magnitudes. The absolute magnitude (M) is the apparent magnitude a star would have if it were brought from its present distance to 10 parsecs (32.6 light years). Consider the following example. A star at 40 parsecs is 4 times farther away that at 10 parsecs. If brought to 10 parsecs, it would be 4 times closer, and by the inverse square law of brightness changes, it would be 16 times brighter. (The inverse square law of brightness means that if the distance of a source of light is *increased* by 2, 3, or 4 times its present distance from the observer, the brightness of the source will *decrease*, respectively, by 4, 9, or 16 times. The reverse will be true if the source is brought closer to the observer.) This is nearly the same as $2.512 \times 2.512 \times 2.512$, which means that the star would be 3 magnitudes brighter. If the star had an m of $+9$ at 40 parsecs, its apparent magnitude at 10 parsecs would become $+6$, and by definition its *M* would be +6. The difference *m* − *M* is 3. This number is called the **distance modulus**, which means that it is the number of magnitudes by which a star becomes brighter when it is brought from its present distance to 10 parsecs. In effect, the distance modulus is just another way of expressing the distance of a star.

Consider another example. Let the distance modulus of a star be 10, which means that the star is 10 magnitudes fainter at its present distance than at a distance of 10 parsecs. Ten magnitudes may be written as 5 + 5; since each 5 magnitudes corresponds to a ratio of brightness of 100 times, the 10-magnitude difference is equal to a brightness ratio of 100 × 100, or 10,000 times. But, by the inverse square law, a change of brightness of 10,000 times is the result of a distance change of 100 times, or the square root of 10,000. Therefore, the star is 100 times farther away than 10 parsecs, or its distance is 1000 parsecs (3260 light years). We shall use the idea of distance modulus a number of times in later discussions.

6.16 STELLAR SPECTRA AND THE SPECTRAL SEQUENCE

The spectra of about a half-million stars have been photographed. Almost all stellar spectra are of the absorption type, although a few have emission lines. After a great deal of early study, it was found that almost all stellar spectra could be placed in a sequence called the **spectral sequence**, in which there are seven main classes designated O, B, A, F, G, K, and M. Except for class types O and M, there are ten subdivisions in each class. With the exception of type O, representative samples are given in Figure 6.5. One of the greatest discoveries in astronomy was the revelation that the spectral sequence is really a **temperature sequence**. The type O stars are the hottest, and those of type M are the coolest. A brief description of each type follows.

Type O   Surface temperature 50,000K. The few lines are due to ionized silicon, helium, oxygen, and nitrogen. H (hydrogen) lines very weak. Color blue.

Type B   Temperature 16,000K. Neutral helium but none ionized. Singly ionized oxygen and silicon. H lines stronger. Color blue-white.

Type A   Temperature 9000K. Strong, broad H lines. Singly ionized magnesium and faint lines of neutral metals begin to appear. Color white.

Type F   Temperature 7000K. H lines weaker than in type A, but still strong. Numerous lines of neutral metals chromium, iron, and calcium, but singly ionized metals still show faintly. Color yellow-white.

Type G   Temperature 5500K. H lines continue to weaken. Most lines are of neutral atoms. Singly ionized calcium strong, but neutral calcium weak. Color yellow. Sun is type G2.

Type K   Temperature 4500K. Most lines from neutral metals. H lines very faint. Singly ionized calcium strongest, but neutral calcium stronger than in type G. Absorption bands of molecules begin to appear. Color orange to red.

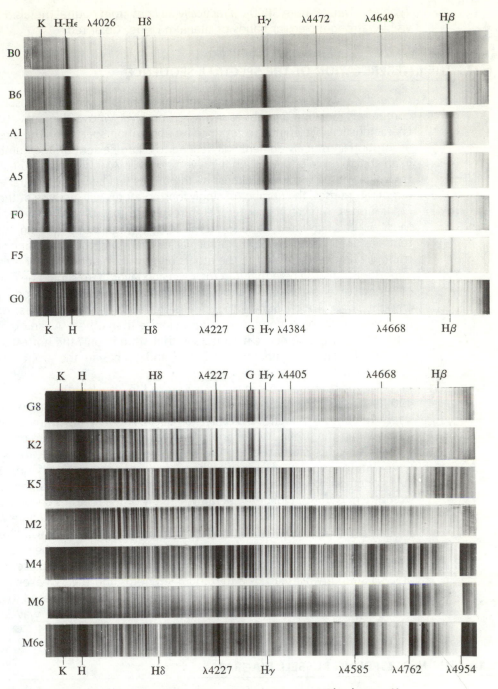

FIGURE 6.5 Some representative examples in the spectral sequence. The letters K and H refer to the lines of singly ionized calcium. The letter H followed by Greek letters marks hydrogen lines; λ4227 is a line of neutral calcium, and λ4668 is due to helium. (Photo from University of Michigan Department of Astronomy.)

Type M   Temperature 3000K. Practically all lines from neutral metals. H lines barely visible. Strong molecular bands of titanium oxide. Color red.

6.17   THE INTERPRETATION OF THE SPECTRAL SEQUENCE

Generally, most stars are constructed much like the sun, with a photosphere emitting continuous radiation and above it a reversing layer whose elements absorb at their characteristic wavelengths to produce an absorption spectrum. The strength of the absorption lines of an element will depend to a large extent on the number of atoms in an absorbing column through the reversing layer. At low temperatures, the thermal velocities will be small, and almost all the atoms will be in the neutral state. As the temperature rises, the number of singly ionized atoms will increase at the expense of neutral ones. At even higher temperatures, there may be almost no neutral atoms. Most will be singly ionized and some will be doubly ionized, and so on, toward higher temperatures. Therefore, at any particular temperature, the appearance of the spectrum will be directly related to the atoms present in the neutral or ionized state. It is important to realize here that if in the reversing layer there should be equal numbers of atoms of two separate chemical elements, they need not produce the same amount of absorption. Some atoms are better absorbers than others. Again it must be realized that some atoms are harder to ionize than others. As an example, helium is hard to ionize, and hence it appears in that state in only the hottest stars. On the other hand, calcium is rather easy to ionize, and its lines are the strongest in the much cooler stars.

The H absorption lines in Figure 6.5 are in the wavelength range from 3900 Å to 4900 Å. Those that appear in this range arise from transitions from the second energy level in that atom to higher levels (Section 3.13). At low temperatures, most of the H atoms are in the ground state, and only a few are available for transition from the second to higher states. Hence, the H lines in the above range are weak. As the temperature rises, the population of the second level increases, and more H atoms are in a condition to absorb from the second level to higher levels; hence, the lines strengthen. At much higher temperatures, many of the H atoms become ionized and, therefore, fewer are available to absorb. As a result, the lines become weak again.

In Figure 6.5, observe the K line of singly ionized calcium (Ca II) and the line 4227 Å of neutral calcium (Ca I). The neutral line is strong in the cool stars, but it weakens as the temperature increases and more ionization takes place. Then the line of Ca II strengthens, reaches a maximum strength, and then weakens again in the very hottest stars, as the number of singly ionized atoms is decreased to produce the doubly ionized type for which no absorption lines occur in the wavelength range shown.

6.18   THE HERTZSPRUNG-RUSSELL DIAGRAM

Shortly after 1910, the American astronomer Henry Norris Russell and the Danish astronomer Ejnar Hertzsprung discovered a powerful correlation between spectral type and absolute magnitude. When the absolute magnitude (*M*) and the spectral types of many stars are plotted together in Figure 6.6, two main features are

FIGURE 6.6 The Hertzsprung-Russell diagram.

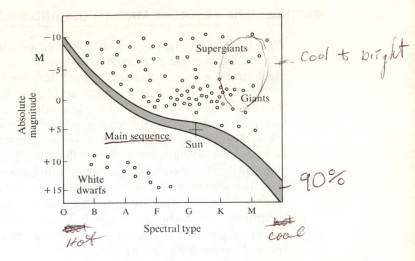

cool to bright

90%

cool Hot

cool

observed. The great majority of stars for which data are available lie in a narrow band called the *main sequence*, which runs across the diagram from about $M = -10$, for the hottest type O stars, to about $M = +15$, for the coolest type M stars. Most of the other stars, from types G through M, are in another broad band above the main sequence in a region of the giants and supergiants. The position of the sun is marked with a cross on the main sequence. From one end of the main sequence to the other, there is an extreme difference of 25 magnitudes, corresponding to a luminosity ratio of 10 billion to 1. Note that at a type such as K there is a difference of several magnitudes between a giant and a main sequence star of the same type. Since the energy flux per unit area and unit time depends on the fourth power of the absolute temperature (which is the same for both stars), the K giant must have a much larger surface area and diameter than the main sequence K star. The quoted luminosity ratio of 10 billion to 1 between the hottest O star and the coolest M type does not represent the ratio of surface areas because the O star is much hotter and puts out a great deal more light per unit area and time than does the main sequence M.

Well below the main sequence is another less conspicuous sequence called the *white dwarfs*. The adjective "white" comes from the fact that the first of these to be discovered, the faint companion of the bright star Sirius (alpha Canis Majoris), is white. The adjective is still used even though some white dwarfs are yellow. A particular white dwarf, say, of type F, will be several magnitudes below the main sequence. Since an F star on the main sequence and the type F white dwarf have the same surface temperature, the white dwarfs must be much smaller. Many are about the size of the earth. However, their mass is about equal to that of our sun, and from this we conclude that the density of the white dwarf material must be exceedingly high. Densities of up to 100,000 times that of water are not uncommon. A cubic inch of such material would weigh not quite two tons! This unusual state of matter may be partly explained by saying that most of the atoms are stripped bare of their electrons right down to the nucleus, thus allowing the gas to be highly compressed. It appears that white dwarfs are dying stars and that they are the last stage in the evolution of a star. They have lost almost all of their hydrogen and other nuclear fuels, so they shine only by their internal heat.

6.19 SPECTROSCOPIC ABSOLUTE MAGNITUDES AND DISTANCES

Within certain limits, the H-R diagram can be used to determine stellar distances. Assume that the star is on the main sequence and that its spectral type and apparent magnitude are known. Reading upward from the spectral type to the main sequence and then to the left, one obtains the absolute magnitude. The distance modulus and then the distance follows from the absolute magnitude M and its apparent magnitude m. For stars not on the main sequence, the method is somewhat complicated by the vertical spread in M at each spectral type among the giants and supergiants. However, the absolute magnitude in these regions can be obtained for each spectral class (surface temperature) by a comparison between the intensities of those spectral lines that are sensitive to luminosity and others that are not. An examination of the spectral lines will show whether the star is a main sequence star, a giant, or a supergiant. This technique makes it possible to find the distance of a very distant giant or supergiant as long as one can obtain its spectrum and classify it for both its spectral type and luminosity class.

6.20 THE MASS-LUMINOSITY RELATION AND THE LUMINOSITY FUNCTION

The mass-luminosity relation expresses the fact that for the majority of stars, the more massive the star, the greater is its luminosity. The luminosity increases approximately as the power 3.5 of the mass, which means that a small increase in mass produces a large increase in brightness. The relation holds best for stars on the main sequence of the H-R diagram, fairly well for the giants, and not at all for the white dwarfs. The last are much less luminous than their masses would require them to be.

The luminosity function relates the numbers of stars at each value of the absolute magnitude. The relation shows that the number of stars increases very rapidly as one goes to fainter objects. In a rough way, and to illustrate the extremes involved, there are about 450,000 stars of absolute magnitude +15 for every one of absolute magnitude −5. The function is not too well known because of the problem of statistical sampling. To get a good idea of the number of stars of absolute magnitude −5, it is necessary to use enormous volumes of space because these stars are so rare. At the other end of the scale, the intrinsically faint stars are also apparently faint even when they are close, so that even if one uses a large volume of space they will not be detected because they will be too faint at even moderate distances.

KEY TERMS

photosphere	penumbra	radial velocity
chromosphere	prominences	proper motion
reversing layer	solar wind	stellar magnitudes
prominences	sunspot cycle	absolute magnitude
corona	Maunder Minimum	distance modulus
granules	constellations	stellar spectra

THE SUN: AN INTRODUCTION TO THE STARS

sunspots

umbra

stellar distance

parallax

H-R diagram

QUESTIONS

1. Would it be possible for a planet to revolve about the sun just above the photosphere?

2. What are the physical differences between our sun and a type O star? A type M dwarf? A type F white dwarf?

3. If a star with a surface temperature hotter than that of the sun has the same absolute magnitude in the yellow-green region, will it be brighter or fainter in the ultraviolet region?

4. Why is it not possible to see the solar corona from the earth's surface except at the time of a total solar eclipse or with a coronagraph?

5. What is the ratio of brightness of two stars whose magnitude difference is 12?

6. Discuss the appearance of the sun as seen by an observer on Pluto.

7. If an observer on Pluto were to see a solar flare, how much time would have elapsed between the occurrence of the flare and the observation on Pluto?

8. Both ultraviolet light and the protons of the solar wind may cause aurorae. Upon the outburst of a flare, which type of aurora would you expect to see first and why?

7 Stellar Systems and Variety among Stars

In this chapter, we shall discuss the extraordinary variety found among stars. As we have seen, stars differ enormously in temperature, diameter, color, mass, and luminosity. Some stars, like the sun, show no detectable evidence of variation in brightness, while others may change by many magnitudes in a day or two. There are stars that rotate rapidly and others that seem not to rotate at all. Some systems are double or triple and many may be in a cluster of hundreds of thousands of stars or even a galaxy made up of trillions of stars.

7.1 VISUAL DOUBLE STARS

There are about 50,000 known double star systems whose duplicity can be seen visually with a telescope. The reality of the physical connection of the shorter-period systems is revealed by the revolution of the two stars around each other, just as the earth and moon would be seen to do if viewed from another planet. (See Figure 7.1.) The orbital periods of the visual double stars range from the shortest of 1.7 years to values that may be millions of years. The very long-period systems change their relative positions so slowly that they are known only from their common proper motion.

Because double stars are gravitationally connected, they revolve around each other in an orbit which has a period, an eccentricity, a semimajor axis, and an inclination. Because the orbital planes are randomly oriented in space, the orbital motion of one star with respect to the other will be observed in projection on the plane of the sky. The change in apparent separation and position angle can be plotted after one whole revolution, and from this plot one can determine the true shape of the orbit and its semimajor axis in arc-seconds. The latter, if divided by the parallax, will give the semimajor axis in astronomical units. A proper study of the plot will also reveal the true orbital eccentricity, the inclination, and the spatial orientation of the major axis.

The most important reason for the study of visual binaries is that they are our main source of stellar masses. Kepler's third law in the form given by Newton states

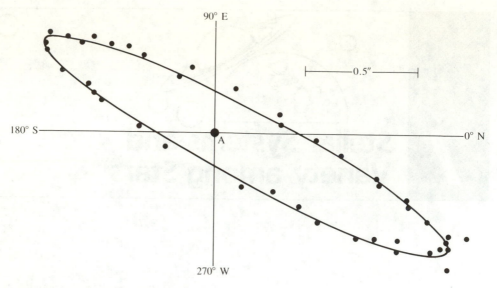

FIGURE 7.1 A plot of the positions of star B with respect to the brighter star A in the visual binary system ADS 10598. The data cover the interval from 1829 to 1950 and the period is nearly 46 years. Although the true orbit is not far from being circular, the projected (apparent) orbit is highly elliptical because of an orbital inclination of nearly 81 degrees. (Data and figure by permission of the authors R. L. Duncombe and Joseph Ashbrook from *Astronomical Journal*, Vol. 57, no. 1199, 1952.)

$$a_{AU}^3 / P_{yr}^2 = m_1 + m_2$$

where the unit of mass is that of the sun. When a and P are substituted in this equation, the result is the mass sum, but not the individual masses. To separate the masses, it is necessary to know the mass ratio m_1/m_2. This ratio is equal to d_2/d_1, where d_1 and d_2 are the respective distances of the center of each star from the center of mass. The position of the latter can be obtained from proper motion studies (see Section 6.13) of the binary. The mass center will move in a straight line with constant proper motion, but the two stars will each revolve about the mass center in a sinuous curve back and forth on either side of the mass center line. The difficulty in making these measurements is one reason why so few stellar masses have been determined.

To accumulate the necessary data on a visual binary with a period as long as 100 years, three or four generations of astronomers must observe the object. This is the kind of cooperation that makes some kinds of astronomical research possible.

7.2 SPECTROSCOPIC BINARIES

For a given binary, if one were to keep the mass sum constant and decrease a, then P would also decrease in such a way as to keep a^3/P^2 constant. The result will be that as both a and P decrease, the orbital velocity of one star with respect to the other will increase. But also, as a decreases, it will be harder to distinguish the system as a visual binary. However, because the orbital velocity increases, the Doppler effect (see Section 3.14) can be used to observe *periodic* changes in the *radial velocity* of each star due to the variable projection of the orbital velocity on to the line of sight to the observer on the earth. Such an object would be called a **spectroscopic binary**. Note that here the observed radial velocity is the projection on to the line of sight of the orbital velocity of each star with respect to the *center of mass*. If the orbital plane were perpendicular to the line of sight, no radial velocity would be observed no matter how large the orbital velocities. The inclination would be defined as zero degrees. As the inclination increases, each star moves back and forth to and from the observer. When the radial velocity changes become large enough to be detected, the object will be listed as a spectroscopic binary. In between zero orbital inclination and close to ninety degrees, radial velocity changes will be observed, but there will be no way to determine the inclination. This can be done only if the latter is so close to ninety degrees that the stars in the pair eclipse each other.

If for simplicity one assumes that the inclination is 90°, one can easily see that the range in radial velocity of each star reaches its maximum. Thus, for each star the total range in radial velocity divided by two is the orbital velocity of each star with respect to the mass center. Assuming that the period is known, the product of the period and the orbital velocity of each star gives the orbital circumferences and hence the radii (for circular orbits), or the distance of each star from the mass center. When inserted into the formula for Kepler's third law (see Section 7.1 and 1.14), the sum of these two radii (expressed in AU) and the period will give the mass sum ($m_1 + m_2$). Again, as in Section 7.1, in order to obtain the individual masses, one needs the mass ratio. In this case the determination is easy because it is nothing more than the ratio of the individual orbital velocities with respect to the mass center. This is true because, according to Section 7.1, a less massive star has the larger orbit and must therefore have the greater orbital velocity in order to make one circuit in the same time as the more massive one. With the mass sum and the mass ratio one can determine the individual mass. This quantity is one of the most important properties of a star.

For most spectroscopic binaries, however, one sees only one spectrum because the other star is too faint to register. As a result, no masses can be determined. This is the state of affairs for the majority of spectroscopic binaries, of which there are several hundred. The total number of well-determined stellar masses is less than one hundred. A proper treatment of the statistics on spectroscopic binaries shows that about half of all stars are binaries.

7.3 ECLIPSING BINARIES

From what has been said in Section 7.2, the binary will be an **eclipsing** one when the inclination is close to 90°, and there will be two eclipses per revolution at the times when both stars are on the line joining the mass center to the observer. From measures of brightness throughout a complete revolution, a light curve is obtained which, when properly analyzed, can reveal a great deal about the physical properties of the stars.

To illustrate, consider the light curve in the lower part of Figure 7.2. This schematic curve was drawn on the assumption that the orbital inclination is 90° (central eclipses), that star B is three times the diameter of A with nine times the

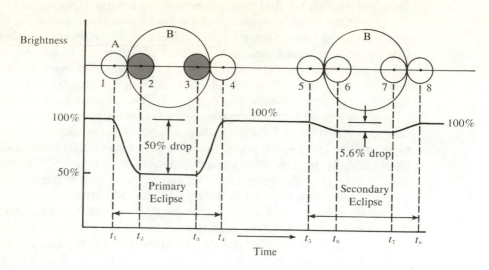

FIGURE 7.2 The contacts and the light curve of an eclipsing binary.

projected surface area, and that each star contributes half of the light of the system. Let us assume that star A passes behind star B and that A moves from left to right while B is fixed. At t_1 star A is just about to begin to pass behind B, while at t_2 star A is completely eclipsed so that the system's light has dropped by 50 percent. The reverse takes place between times t_3 and t_4. In the interval from t_2 to t_3 star A is totally eclipsed. For circular orbits it should be clear that $t_2 - t_1 = t_4 - t_3$ and that each is the time required for the edge of B to cross the diameter of A. Assuming that V is the known orbital velocity of A in its circular orbit around B (considered as the center of the system), then V multiplied by either of the two time intervals gives the diameter of A in miles or kilometers, depending on the units used. Inspection should show that $t_3 - t_1 = t_4 - t_2$ and that each interval is that required for the leading edge of A to cross the diameter of B. Thus, either of these intervals multiplied by V gives the linear diameter of B, the larger star. In order to obtain V, one must see the spectra of both stars and determine the orbital velocity of each with respect to the mass center; V is just the sum of these two orbital velocities. If the orbital velocities are known, one can determine the individual masses (as discussed in Section 7.2), the diameter and volume of each star, the average mass density of each star, and their relative brightness.

In Figure 7.2 positions 5, 6, 7, and 8 and their corresponding times refer to the situation half a period later during which the smaller star passes in front of the larger one. Since A has only 1/9 of the area of B, the drop in light at _secondary_ minimum is only 1/9 of 50 percent, or 5.6 percent, which is small compared with the 50 percent drop at primary minimum. The time intervals will be the same as at primary minimum. As the inclination departs from 90°, the period of totality will shorten until finally there is no flat part at primary minimum and both eclipses are _partial_. Finally, there will be no eclipses at all and the system will be only a spectroscopic binary of unknown inclination. About 2500 stars are known to be eclipsing systems.

FIGURE 7.3 A photograph of a galactic cluster, the Pleiades, in the constellation Taurus. The cluster is imbedded in a dusty region of the Milky Way. Around some of the brightest stars are hairy patches due to starlight reflected by the dust (see Section 7.13). The fan of spikes around the brighter stars is not real, but is an artifact produced by the four-point support of the secondary mirror in the reflecting telescope that took the photograph. (Lick Observatory photograph.)

7.4 MILKY WAY CLUSTERS

There have been discovered about 1100 star clusters of a type found almost exclusively in the Milky Way. These are multiple star systems that are held together by their mutual gravitational attraction. Some are poor, rather inconspicuous objects with as few as 20 stars that barely stand out against the stellar background. At the other end of the scale are a few very rich clusters with more than 1000 stars that stand out in startling contrast. The Milky Way cluster most easily seen by the naked eye is the Pleiades in the winter sky. A search of the Milky Way with a field glass will reveal quite a few of these clusters. Their diameters range from about 15 to 40 light years. These objects, which are also called **galactic** clusters for a reason to be seen later, may be as distant as 15,000 light years—at such distances the angular diameter of the cluster is so small and its stars so faint that the object merges into the rich Milky Way background. (See Figure 7.3.)

7.5 GLOBULAR CLUSTERS

A cluster of an entirely different type is called a **globular cluster** because of its compact, round appearance. About 130 are known. They are found all over the sky and do not favor the Milky Way. Remarkably, the number of cluster members ranges from a few thousand to perhaps as much as one million. When seen in a telescope or on a photographic plate, the star images near the cluster center are so tightly packed together that they cannot be seen as individuals. (See Figure 7.4.) The diameters of globular clusters range from about 75 light years to perhaps as much as 400. Their distances from the sun range from a few thousand light years to over 300,000. The determination of their distances will be discussed in a later section.

It is clear that there is a great difference between these two types of clusters. Their origins and past histories must be quite different. These topics will be discussed later in this chapter. (See Figure 7.7.)

FIGURE 7.4 A 200-inch telescope photograph of the globular cluster M3 in the constellation Canes Venatici. (Palomar Observatory photograph.)

7.6 INTRINSICALLY VARIABLE STARS

Tens of thousands of stars have been discovered that are intrinsically (really) variable. It appears that in almost all cases their light changes are the result of changes in size and temperature. The change in light may be as small as can be detected or up to many magnitudes. For some of these variables, the light changes are closely periodic and of constant brightness range, but for others the changes are quite erratic. This is a rich field for research. This discussion will be limited to two types, the cepheid variables and the novae.

7.7 CEPHEID VARIABLES

Cepheids (seph'-ee-ids) are pulsating stars whose light variations result from a change in surface area and surface temperature. There is actually a change in spectral type during each cycle of variation. For the great majority of cepheids, the time from one light maximum to the next is closely constant, although in a very few well-observed cases, slow period changes have been detected. To a large extent, the shape of the light curve and the range of brightness are the same from one cycle to the next. The two main classes of cepheids are discussed below.

Classical cepheids have periods that range from about 2 to about 100 days. A typical period is about a week, and the typical light range is about one magnitude. A typical spectral type is from G2 at minimum to F2 at maximum. Radial velocity observations confirm the pulsation. The mean absolute magnitude extends over the range from −2 to −6. These are luminous giants. The change in radius is of the order of 5 to 10 percent.

Short-period cepheids have some resemblances to the classical cepheids except that their periods are shorter, their brightness range is less, and they are intrinsically fainter. They are also called **cluster** cepheids because they are often found in globular clusters, although not in Milky Way clusters. However, a few Milky Way clusters do have one or two classical cepheids among their member stars. The range of period for the cluster cepheids is from 0.1 day to nearly one day. The average magnitude range is a little less than one magnitude. Their spectral type is around A and F, and their mean absolute magnitude is close to +0.5.

7.8 THE PERIOD-LUMINOSITY RELATION

One of the most useful discoveries in astronomy was made in 1917 when it was observed that there is a good correlation between the periods and the absolute magnitudes of the classical cepheids. This relationship is called the period-luminosity relation, and it is illustrated in Figure 7.5. In this figure, the absolute magnitude is plotted against the period on a logarithmic scale. Observe that the classical cepheids are confined to a rather narrow band and the longer the period, the brighter the variable. The short-period cepheids are also plotted on this diagram, but there is no change in absolute magnitude with period. Note that the cluster cepheids continue on the diagram into a band called Population II cepheids, and not on into the band of the classical cepheids, which are of type Population I. The meaning of "population" will be discussed at the end of this chapter.

This relation can be used to find the distance of a cluster of stars if one of its members is either kind of cepheid. If it is a classical cepheid, it is necessary to determine first its period and its average apparent magnitude by observation. From the period-luminosity relation, one obtains the average absolute magnitude and then the distance modulus $m - M$, which then gives the distance as described in Section 6.15. If the variable is identified as a short-period cepheid, we need only to know that its absolute magnitude is +0.5; we can then proceed in the same way. The variable can be put in its proper class from the shape of the light curve and the knowledge that the short-period cepheids have periods less than a day and the classical type have periods longer than one day. The cluster cepheids in the globular clusters have given the distances of these clusters. Since the classical cepheids are much brighter objects, they may be used to determine far greater distances, such as those of galaxies. (Galaxies will be treated in the next chapter.)

FIGURE 7.5 The period-luminosity relation for the cepheid variable stars.

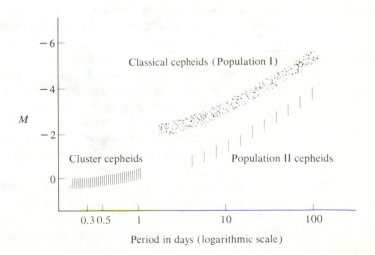

Period in days (logarithmic scale)

7.9 THE NOVAE: ORDINARY AND SUPER

A **nova** (new star) is the most startling of all variable stars. It is a star which without warning increases its brightness many magnitudes in a few days, and then slowly (and sometimes irregularly) fades away to about its preoutburst brightness. The period of fading may require several years. The average *ordinary* nova will increase its brightness about ten magnitudes in three or four days. The decrease is often accompanied by rapid fluctuations which may be periodic for a time. The average ordinary nova will reach an M of -6 to -8 (a very luminous object). An ordinary nova is a rare event. Statistical studies show that 20 to 30 occur in our galaxy per year. Even more rarely is such an object near enough to reach naked eye visibility, and most are found on photographic plates taken for other purposes.

Spectroscopic observations during the outburst reveal that there is a very rapid expansion of the star's outer layers. Expansion velocities as high as 2400 km/s (1500 mi/s) have been observed; these must in every case be greater than the escape velocity from the star. The whole star is not blowing up, but only an outer shell, and the mass loss is probably not greater than about 0.1 percent. There is some evidence that some novae repeat this process after an interval of a decade or so. It has been observed that within some months or years after the outburst of an ordinary nova, a shell or ring sometimes forms around the faded image of the nova. This appears to be the nova's shell of ejected material which has grown to such an angular diameter as to become visible.

A **supernova** is an exceptionally violent phenomenon which is much more rare than an ordinary nova. The increase in brightness may be as much as 20 magnitudes. The absolute magnitude at maximum brightness averages about -17, but it ranges from -14 to -20. Supernovae are often observed in other galaxies, and studies show that in an average galaxy they occur perhaps once every few hundred years. The mass loss in such an explosion is at least 10 percent, and in some cases it may be much more.

7.10 THE CRAB NEBULA

No supernovae have been observed in our celestial neighborhood during the time of modern astronomical instruments, but there are old records of a few. In the constellation Taurus is found the Crab Nebula, a diffuse, bright nebulosity shown in Figure 7.6. This nebula is known to be expanding. From its expansion rate and its present size, astronomers computed (earlier in this century) that the expansion began about the year A.D. 1050. Old records revealed a Chinese source that reported the appearance in A.D. 1054 of an exceedingly bright new star very nearly in that part of the sky now occupied by the Crab Nebula. Astronomers are convinced that the two objects are the same and that the star observed in China was a supernova. The Crab Nebula is a strong emitter of radio noise. Several other galactic nebulae are thought to be the remnants of supernova explosions.

A star near the center of this nebula has been identified as the remnant of the supernova. It varies in brightness in ordinary light in a period of only 0.033 second. It also emits short pulses of radio noise with the same period. This type of object is called a *pulsar* and will be described in more detail later in this chapter. This object in

FIGURE 7.6 The Crab Nebula in the constellation Taurus. The photograph shows the distribution of glowing hydrogen gas (Lick Observatory photograph.)

the Crab Nebula is a strong emitter of X-rays, whose partial absorption by the nebular gas probably contributes to its continued brightness.

7.11 OUR GALAXY: THE MILKY WAY

Space Distribution

Galileo first observed that the number of stars seen in the field of view of his telescope when he was looking at the Milky Way was much greater than in directions away from it. This simple observation made more than 3½ centuries ago has had a great influence on the course of astronomy. This narrow band of light (the street of milk) around the sky was revealed to be a zone of great numbers of stars too faint to be seen as individuals by the unaided eye. When the sky is viewed at right angles to the Milky Way and one looks at the poles, relatively few stars are seen. When the telescope is moved along the Milky Way, the number of stars in the view field fluctuates a good deal. However, the richest region is seen in the constellation Sagittarius in the southern summer sky, and the least rich is in that part of the band 180° away in the northern sky. Our solar system is a part of this galaxy.

The interpretation of these observations is that our sun (the place of observation) is in a large flattened cluster of stars and that our position is considerably off-center. This system is called our **galaxy**, or the home galaxy. There are many galaxies outside our own (see Chapter 8). Figure 7.7 is a schematic view of our galaxy seen from well outside and on edge. The sun's distance from the center is about 28,000 light years. The vertical thickness of the disc at the position of the sun is about 3000 light years. The lens-shaped figure with the **central bulge** represents the envelope within which are found most of the **luminous** stars in our galaxy. The diameter of this feature is about 100,000 light years, and it contains about 200 billion stars. This latter number is also about its mass in solar masses. Seen from above or below, it would be roughly circular in outline, with spiral arms trailing outward from a region about

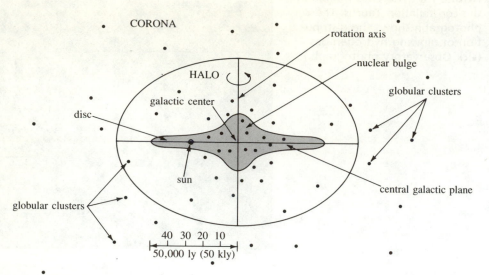

FIGURE 7.7 A schematic view of our galaxy seen on edge showing the position of our sun and the subdivisions of the galaxy.

15,000 light years from the center. In the spiral arms of the thin disc are found the Milky Way clusters, stars of a wide range of brightness, and clouds of gas and dust. Here in the disc most of the stars are relatively young. In the central nuclear bulge most of the stars are red and old.

The globular clusters are found mostly within about 20,000 light years of the galactic center. This system of clusters has the same center as that of the disc and the nuclear bulge. The system is somewhat flattened in the direction of the rotation axis of the disc and bulge; this may indicate some rotation of the whole globular cluster system. See figures in Chapter 8 for examples of galaxies similar to our own.

Outside the limits of the nuclear bulge and the disc there is another space called the **halo** which, in the plane of the disc, has a radius of about 65,000 light years. It is somewhat flattened in the direction of rotation of the disc. In this space one finds mainly old stars such as cluster-type cepheid variables, white dwarfs, a few globular clusters and numerous other types. In the disc the great majority of the stars are revolving in nearly circular orbits about the center with only quite small inclinations. But in the halo the stars tend rather strongly to revolve about the galactic center in fairly eccentric orbits with rather high inclinations to the galactic plane.

Beyond the boundary of the halo is a fairly new space called the **corona** with a rough outer limit of perhaps 200,000 light years in all directions from the galactic center. As we shall see, in the section entitled "The Mass of Our Galaxy," the corona may well contain a major part of the total mass of our whole galaxy.

The Motions of Galactic and Globular Clusters

Galactic clusters are almost all found very close to the galactic plane and revolve about the galactic center in near circular orbits. The star density (stars/cubic light year) in the cluster is much smaller in galactic clusters than in the globular variety, where it can be very large near the cluster center. A galactic cluster is held together by the mutual gravitational attraction of its members. If this were not so, the cluster would rapidly disintegrate after its original formation and disperse. However, galactic clusters do have a finite lifetime. One reason is that the gravitational attraction of the massive nuclear bulge wants to make the cluster members nearest to the galactic center have orbital velocities greater than those farther away (Kepler's third law). If this were to prevail over the gravity in the cluster, then, in a short time on the cosmic time scale, the cluster would elongate into a sausage-shaped configuration and finally disperse altogether in about one complete revolution about the galactic center. Mutual gravitation tends to prevent this, but not altogether. As a result of internal motions in the cluster, members are gradually lost; in anywhere from a few hundred million to a few billion years the cluster is no more. The clusters with the greater mass density and greater mass will last the longest. The process is accelerated somewhat by the passage through the cluster of nonmembers which make close gravitational encounters with cluster members and give them the energy to escape.

Globular clusters have much longer lives because of their much larger mean density and mass and the greater gravitational attraction of the members for one another, and also because they spend most of their lives in relative isolation from other stars. Globular clusters, wherever they are, are thought to revolve in fairly elliptical orbits around the galactic center. Thus a globular cluster from far above or below the galactic plane will, in its revolution around the center, pass through the dense star regions of the galactic plane once per orbital period. In the process there will be a gravitational collision (close approach, not physical contact) with single stars and tidal distortions caused by the central nuclear mass. One might expect that globular clusters that spend their time close to the nucleus in small galactic orbits would have lost more members; indeed, it is observed that the inner clusters have smaller masses than the far remote ones that spend most of their existence in remote isolation. While galactic clusters in the disc are often observed imbedded in clouds of gas and dust in the spiral arms, globular clusters show no evidence of the presence of this interstellar medium. If any was ever present, it was probably swept out after one or more passages through the galactic plane. Possibly some of the dust and gas was absorbed by the stars in the cluster.

The Mass of Our Galaxy

The determination of the mass of our galaxy would be a relatively simple matter if we knew the distance to the center, had a good value for the sun's circular orbital speed around the galactic center with respect to a frame of reference at rest, and could assume that the great majority of the galactic mass was inside the sun's distance from the center. Determination of the mass was first tried by assuming that the nearly spherical globular system was not rotating. Radial velocity measures of the globular clusters showed that our sun had an orbital velocity of about 180 km/s (110 mi/s) with respect to the globular cluster system as a whole, if the system is assumed to be at rest and not rotating. In search for what might be a better standard of rest, astronomers measured the sun's motion with respect to a fairly numerous group of external galaxies and found a value of around 250 km/s (150 mi/s), which suggests that the globular cluster system is rotating and should not be used as a reference system. With this new orbital velocity and adopting a distance to the galactic center of 28,000 light years, one can compute the galactic orbital period of the sun to be 200 million years. Now using the second form of Kepler's third law (and the proper units), one computes that the mass inside our sun's galactic orbit is 120 billion solar masses. However, the use of Kepler's third law is valid only if the mass in all directions outside our sun's distance from the galactic center is negligible. In our solar system the mass of the sun is 500 times that of all the planets put together. Therefore, the third law will rather accurately predict the decreasing planetary orbital velocities with increasing distance from the sun. But in recent years refined measurements have shown that the galactic orbital velocities of a number of objects at greater distances from the center than our sun are *increasing* with distance and that at about 55,000 light years from the center the speed rises to 300 km/s (180 mi/s). Therefore, there must be a great deal more mass outside our distance from the center. Careful studies show now that the total mass of our galaxy is at least one trillion solar masses and probably more. If this is correct, then why do we not receive light from this large additional mass? One suggestion is that in the outer corona of our galaxy there are very large numbers of old, very low-mass, extremely underluminous objects which we might not even call stars. Please reread Section 6.20 to recall that as the mass of a starlike object decreases its brightness decreases at a far more rapid rate. It is not out of the question that many of these highly underluminous objects are like the planet Jupiter, whose mass is so small that, if it were not for the sun's light, would be invisible. Therefore, it appears that the mass of our galaxy may be as much as ten times greater than it was thought to be ten or so years ago; this may apply to many other galaxies.

7.12 THE INTERSTELLAR MEDIUM

Although the patchiness of the Milky Way can be observed by the naked eye, the detailed structure is revealed only by large-scale photographs, as in Figure 7.8. A typical photograph will show great differences in the number of stars per square degree. This nonuniformity is caused almost entirely by huge clouds of interstellar dust and gas. This material is very strongly concentrated toward the plane of the Milky Way (the galactic plane) and in the spiral arms. All the space in our galaxy is pervaded by this medium, but there are large variations in its density. The more obvious, obscuring patches result from the larger, closer, and denser clouds, some of which reduce the brightness of the stars behind them by as much as 30 magnitudes. See Figure 7.8.

Most of the gas between the much denser clouds consists of hydrogen atoms with a density of about one per cubic centimeter, but other chemical elements are also present in much smaller numbers. For about every 1000 hydrogen atoms, there is probably one dust particle with a diameter of 10^{-5} centimeters (1000 Å) or less. The dust and gas occur together and the total mass of the dust is about 1 percent of the whole medium. In spite of the relative scarcity of the dust particles, they are the main contributors to the obscuration of light. The total mass of the interstellar material is much less than that of material in the form of stars. Even in the thickest parts of the clouds, the density is still so low as to constitute a very high vacuum by laboratory standards.

FIGURE 7.8 Nebulosity in the constellation Monoceros. Note that the tongue of dark nebulosity which narrows from right to left towards the center of the photo is converted at its tip into a bright patch of emission nebulosity by a very bright star a little farther to the left. (Palomar Observatory photograph.)

7.13 THE EVIDENCE FOR THE INTERSTELLAR MEDIUM

Interstellar Reddening

It has been known for a long time that star colors become progressively redder for the fainter and more distant stars. The interstellar material (almost exclusively the dust particles) scatters the starlight out of the beam coming to the observer and causes it to appear fainter than it would otherwise be. The scattering is greater for the shorter blue wavelengths than for the red, with the result that a very distant blue type O or type B (see Section 6.16) star may appear red. This reddening is quite slight at a few hundred light years, but it becomes important for distances of much more than 1000 light years. For distances in the galactic plane greater than about 15,000 light years, we can scarcely see any stars because of this obscuration. In fact, our direct view of the galaxy is quite limited and is confined in the galactic plane to our rather immediate neigborhood. As soon as one looks in directions somewhat above or below the galactic plane (i.e., beyond the band of the Milky Way), the obscuration is greatly reduced because the obscuring material is so closely confined to the galactic plane.

Bright Nebulae

Among the darker, absorbing regions are often found bright nebulae of considerable extent which are associated with dark nebulae. These are explained by the presence near to, or even within, the dark material of very hot, bright stars which illuminate a portion of the dark nebulosity. The famous Orion Nebula is a good example. (See also Figure 7.8.) If the illuminating star is very hot, its copious ultraviolet radiation will excite and ionize the gases in the clouds, so the spectrum of the bright nebula is a bright-line (emission) spectrum characteristic of the elements of the gas. If the star is too cool, the excitation will be small, and the spectrum will be much like that of the illuminating star reflected from the dust. (See Figure 7.3.)

Interstellar Absorption Lines

Among the numerous examples of spectroscopic binaries are some with type O and type B components. Although they may be quite distant, they are bright enough to have their spectra recorded photographically. Most of the stellar lines are broad and change their wavelengths with the orbital period, but for the more distant binaries, there are usually found a few narrow absorption lines of *constant* radial velocity. The most common of these are identified with neutral and singly ionized calcium, iron, and sodium. These sharp lines were once thought to be produced in an extended atmosphere around the binary. However, it was found that their strength increased with the distance of the binary. The most obvious explanation is that these lines are produced by absorption in the space between the binary and ourselves. Commonly, these sharp lines are multiple. Each component is produced by a separate cloud, and the wavelength separation of the components is the result of the different radial velocities of the clouds.

The Radio Emission of Interstellar Hydrogen

Just after world War II, it was predicted, and soon observed, that atomic hydrogen in interstellar space is emitting a single emission line with a wavelength of a little more than 21 cm (about 8 in). It is possible to detect neutral hydrogen clouds at great distances in the Milky Way plane and to map their positions even beyond the galactic center to the opposite edge of the galaxy. Neutral, atomic hydrogen is a poor absorber of its own 21-centimeter radiation. As we said previously, observations in the optical region do not extend beyond about 15,000 light years from the sun. The results show that there are great arms of hydrogen clouds spiralling out from the galactic center and that these arms contain the great proportion of the stars in our galaxy. Our sun is in one of the spiral arms.

Interstellar Molecules

Interstellar clouds of atomic hydrogen have been known for a long time as a result of their association with bright, hot stars in the bright nebulae and also from the presence of the 21-centimeter radiation of neutral atomic hydrogen. In the visible region, absorption lines have long been detected for calcium, sodium, iron, and titanium, as well as for the molecular radicals CN and CH^+. With the development of radio astronomy, the spectral lines in the radio region have been discovered for about fifty molecules. Among these are the radical OH, water (H_2O), cyanogen (C_2N_2), ammonia (NH_3), formaldehyde (CHOH), carbon monoxide (CO), cyanoacetylene (HC_3N), methyl alcohol (CH_3OH), and formamide ($HCONH_2$). Molecular hydrogen (H_2) has been detected in the deep ultraviolet from rocket observations above the earth's atmosphere. Most of these molecules appear in sources with small angular diameters and hence small linear diameters. It has been suggested that these molecules are present and forming in the cool outer regions of clouds of dust and gas that are condensing to form a star. Note the similarity between some of these molecules and those found in comets and in the atmospheres of Jupiter and Saturn. Although these are classed as organic molecules, they were certainly not formed by life processes; however, their presence in the atmosphere of proto-earth may have contributed to the beginnings of life on earth and on planets in other systems.

7.14 THE FORMATION OF STARS

Scattered about in many parts of the Milky Way, and particularly in regions of heavy obscuration, are large numbers of small dark globules of dense interstellar matter thought to be in the process of condensing to form stars. Rough measurements suggest for these globules masses sufficient to form a single star, a binary, or even a star cluster. Theory suggests that there is an upper limit to the mass of a star, and none are known with masses more than about 60 times that of our sun. The relative scarcity of very massive stars suggests that smaller masses are preferred. If the mass of the globule is greater than about 60 solar masses, then at least two stars will be formed. Present theory proposes that as the globule becomes smaller under its own gravitational attraction, its temperature increases by the transfer of gravitational energy to the thermal (kinetic) energy of the gas. At an early stage in this process, the star will have a low surface temperature, a very red color, and a faint absolute magnitude which would place it far to the right in the H-R diagram. As contraction proceeds, both surface temperature and luminosity will change until finally the star enters the boundaries of the H-R diagram. From here it will proceed downward and to the left to that point on the main sequence that is appropriate to its mass. Its movement across the H-R diagram, ending on the main sequence, is very rapid. By this time its central temperature is so high that it is able to support the nuclear reactions that are discussed in the next section. At this point the outward gas pressure due to the high internal temperature exactly balances the inward gravitational force, and the star ceases to contract. The star is then said to be in *hydrostatic equilibrium*.

7.15 NUCLEAR REACTIONS IN STARS AND STELLAR EVOLUTION

This is an extensive and important topic which can be given only brief treatment here. The source of the immense energy radiated by stars for so long a time was a great puzzle until, early in this century, Einstein developed his relativity theory and proposed that mass was just another form of energy like heat, light, and electrical energy, among others. The relation between mass and other forms of energy was derived by Einstein in the famous equation $E = m \times c^2$, where m is the amount of annihilated mass, c is the speed of light, and E is the amount of energy obtained from the annihilation of that mass. This process is extremely important; for instance, total annihilation of one pound of coal would produce about *one billion* times as much energy as would be produced if the coal were burned in the usual manner! Astronomers now believe the most important source of stellar energy is the conversion of hydrogen to helium, which can take place only in the central region of a star where the temperature is around 10 million K (see Appendix 2, II). Energy is released because the mass of the four hydrogen atoms is greater by about 0.7 percent than the mass of the helium atom formed; it is this mass excess which appears as energy. In the case of our sun, the total energy radiated per second is equivalent to the annihilation and loss of two million tons per second. Since the sun has a total mass of 2.2×10^{27} tons and about 65 percent of it is hydrogen, one can show that the sun will continue to shine at about its present rate for several more billion years beyond its present age of about 4.5 billion years.

When a star, regardless of its mass, reaches the main sequence on the H-R diagram, it has ceased to contract and its only energy source is the conversion of hydrogen to helium. However, some of the blue supergiants with luminosities greater than one million times that of the sun are consuming their hydrogen just that much faster. It is true that the mass of the blue supergiant is greater, but usually by not much more than 30 to 50 times. One must conclude that these highly luminous objects are evolving and aging very rapidly and that they must be very young in years compared to the sun.

As the star uses up its hydrogen, it slowly leaves the main sequence and moves upward and to the right in the H-R diagram toward the region of the redder giant stars. The rate at which it does so depends on its mass. A type M main sequence dwarf will probably stay on the main sequence for many billions of years before it moves up appreciably; but a type B or type O star would move an appreciable distance upward in a few million years. This can be seen quite easily when the spectral types (or temperatures) and absolute magnitudes of galactic cluster stars are plotted on the H-R diagram. Presumably, all the cluster stars are formed from a large globule at very nearly the same time, and all are of nearly the same age. If the brighter members of the cluster are type O and type B stars, it is seen that they have already left the main sequence and are above it, while the faintest and less massive members much farther down the main sequence fit closely to that sequence.

Theory shows that as the hydrogen becomes exhausted in the central core, the core becomes smaller by gravitational contraction, and the release of gravitational energy increases the core temperature. At the same time, this hotter core causes the outer atmosphere to expand, and the star moves into the yellow-giant and red-giant region above the main sequence. At a very elevated temperature, the core temperature becomes high enough so that helium is converted into heavier elements with the release of nuclear energy. Nuclear theory shows that this fusion process cannot continue beyond iron, at which point there are no more sources of nuclear energy. At some time during one of these phases of core contraction, the star may become unstable and pulsate, showing a variation in brightness. In some instances, the star may become a nova, the type depending upon the star's mass. The actual details of the evolutionary track on the H-R diagram are quite complex and much influenced by the star's mass.

If the star's mass is no more than about 1.5 solar masses after the nova stage, it will continue to contract until it reaches the *white dwarf* stage, at which point it will have a diameter about that of our earth. These stars form a sort of sequence well below the main sequence on the H-R diagram, as shown in Figure 6.6. There are about 250 known white dwarfs. The star contracts because it has used up all its nuclear energy sources and can no longer produce the heat and the gas pressure to balance the inward gravitational attraction. But as the star collapses, gravitational energy is converted into thermal energy, the interior gets hotter, the gas particles move faster, the atoms collide with greater violence, and the gases of all the elements are completely ionized. This leaves a gas consisting of the atomic nuclei and electrons. As the contraction continues the internal temperature rises, and finally the gas pressure (due mainly to the more numerous electrons) reaches a value which matches that of the inward gravitational attraction. The star ceases to contract and is stable. Once again the star is in hydrostatic equilibrium. The white-dwarf stage is unusual in that the average object has about one solar mass packed into a volume about that of the earth. Thus, the average density may easily be 100,000 times that of water as compared to our sun, with an average value of 1.4 times that of water. For this white dwarf, a cubic inch of the substance would weigh about *two tons*! Even though the surface temperature may be quite high, the surface area is relatively very small, so a white dwarf has a large (faint) absolute magnitude and usually a faint apparent magnitude even though it may be fairly close.

When a stellar nucleus collapses to form a white dwarf, it is believed that the greatly increased surface temperature heats the star's outer atmosphere to such a great extent that it explodes outward and becomes an ordinary nova as described in Section 7.9. It is thought, however, that the original violent explosion may be followed at unknown time intervals by one or more very minor events in which shells of gas are ejected at relatively low velocities of 30–50 km/s (20–30 mi/s). After a time the shell may become visible with a telescope and form a disc which resembles a planet to a viewer; these discs are therefore known as **planetary nebulae.** Figure 7.9

FIGURE 7.9 A planetary nebula in the constellation Aquarius. The light of the slowly expanding shell is caused by the excitation of the shell gases by the intense ultraviolet radiation from the very hot star at the center of the hole. (Palomar Observatory photograph.)

is a photograph of such a nebula in which at least one shell was ejected perhaps 40,000 years ago. These planetary nebulae receive the energy for their radiation from the copious ultraviolet radiation of the star at the center of the shell. These central stars are probably white dwarfs with a surface temperature of about 100,000K. Eventually the shell will expand to such a size that its surface brightness will be too small for it to be detected. The lifetime of such stars is probably around 100,000 years. Undoubtedly those that are dying out are being replaced by others in the manner described here. There are about 1500 known planetary nebulae and probably many more whose light is obscured by the interstellar dust. Most planetary nebulae are strongly concentrated toward the galactic center.

Another suggestion for the origin of ordinary novae arises from the fact that the nova may be one of the components of a close binary system and that matter from the larger, more massive star is streaming on to the smaller, less massive star, which is actually a very hot white dwarf. As matter, mostly hydrogen, accumulates on the white dwarf, nuclear reactions begin to convert hydrogen into helium with the rather sudden release of luminous energy, resulting in a nova. This could explain why some novae are repetitive but not necessarily with a simple periodicity or with the same increase in brightness.

7.16 NEUTRON STARS AND PULSARS

For some time it has been predicted that there might be an additional contraction stage after the white dwarf stage if the star were originally more massive. Present theory (with some uncertainty) indicates that if the original stellar mass were between two and eight solar masses and it ran out of nuclear fuel, the collapse might be signalled by one of the fainter types of supernovae. Quite a bit of mass could be lost. This more massive object would collapse rapidly, as in the case of the white dwarf, but because its mass was greater, it would go through the white dwarf stage and continue its collapse with ever-increasing internal temperature. Finally, the collisional speeds of the atomic nuclei would be so great that they would be smashed into protons and neutrons, the electrons would combine with the protons to form neutrons, and the whole object would become a neutron star with a diameter of roughly 16–30 km (10–20 mi). Its mass might be about two solar masses and it would be composed almost entirely of neutrons. The surface gravity and escape velocity for a white dwarf would be thousands of times greater than for a body like our sun. For a neutron star these quantities would be enormously larger.

Until 1968 no one had ever discovered a neutron star, but in that year at Cambridge University, the first **pulsar** was found. To date somewhat more than 100 are known. Broadly speaking, a pulsar is an object which emits short, periodic bursts of radio noise, whose durations are small compared with the interval (period) between bursts. A study of the distribution of energy with wavelength shows that this is synchrotron radiation, which means that this energy is being radiated by very high-speed electrons moving in a strong magnetic field. Pulsar periods range from a few seconds to as little as 0.0015 second (1.5 milliseconds). The pulsar in the Crab Nebula has a period of close to 0.033 second (33 milliseconds)—about 30 rotations per second. This pulsar and one other are the only ones that have been detected optically. It is now believed that a pulsar is a rapidly rotating neutron star with a strong magnetic field which rotates with the star and drags the electrons along with the field at high speeds. A pulsar cannot be a white dwarf because a white dwarf the size of our earth, rotating once per second, would disintegrate due to the gross excess of centrifugal force over gravitation. The same mass crammed into an object 16 to 30 km (10 to 20 mi) in diameter would have the requisite gravitational force to hold the object together even though there might be a sizeable equatorial bulge. The most likely candidate is a neutron star.

Most pulsar models are not very satisfactory, but one of the best is the lighthouse model. Just as a marine lighthouse makes one complete revolution in a given period and emits a narrow bright beam which is seen by a ship at sea for a brief moment, so the pulsar may emit a narrow cone of radiation which, as the pulsar rotates, traces out a narrow band on the sky. If the observer is in the band, a pulse of radiation is observed, but outside the band nothing is detected. If this is correct, then there must be a great many pulsars that we do not detect. The reason why the radiation is emitted in a narrow cone is unknown, but it may be related to a situation in which the axis of rotation and the magnetic axis do not coincide. If the radiation comes out in a narrow cone along the magnetic axis, then it will sweep out a narrow band as suggested above. Then again, there may be a hot spot somewhere on the star consisting of a dense mass of electrons pulled around by the pulsar's powerful magnetic field. If the rotation speed is very high, the synchrotron radiation will be emitted in a narrow cone and will behave as described above. There really is no satisfactory explanation except that the pulsar must be rotating.

When pulsars were first discovered, careful measures showed that the interval between pulses was constant to at least one billionth of a second. Later work showed that the period was getting longer but not gradually so. The period slowly increased for a time and then suddenly changed to a slightly longer period. One explanation is that the cross section through the poles is elliptical and the figure of the object is a balance between the gravitational and centrifugal forces. Since the pulsar radiates and loses energy and because it is thought the pulsar's only energy source is that of rotation, it must follow that for the pulsar to keep on radiating, it must slow down in the process of converting rotational energy into radiant energy. Now if the surface of the pulsar consisted of a frozen crust of neutrons and the pulsar was slowing down, the figure would have to become less elliptical, the crust would break to readjust the figure, the crust would freeze again, and the period would be a little longer. Similar period readjustments will follow as the pulsar slows down. If this explanation is correct, then one could say that there are "starquakes."

One binary pulsar is known in which the pulsar period is 0.059 second and the orbital period about the unknown, but probably dwarf, star is 7.5 hours. This is a most unusual system.

7.17 BLACK HOLES AND X-RAY STARS

Although the observed white dwarfs and the neutron stars are already sufficiently fascinating, theorists have suggested a third stage of collapse—a star with a mass of not less than about ten times that of the sun and which runs out of its nuclear fuel and begins its rapid collapse with a rapidly increasing temperature, probably ejecting a good deal of mass as a supernova. It would probably end as a stable object with a diameter of perhaps 3 km (1.8 mi) and a mass of at least a few solar masses. The actual size of such an object is not known, but it can be shown that there is a distance from the object's center called the *event horizon* where the gravitational attraction is so great that no light or radiation of any kind that originates within this boundary can escape. As a result, the object is forever invisible and has been named a **black hole**. Because light photons have energy, they have mass and as a result are subject to gravitational attraction. For a black hole, the speed of light must be less than the escape speed; hence photons cannot escape at distances from the center of the black hole body greater than the event horizon radius. For escape speed see Section 4.13. The only way to identify a black hole would be to find an invisible object that was part of a binary system. One candidate is an object called Cygnus X-1. The X means that it is a source of X-rays. This object exhibits the spectrum of a large, hot type B star; measures of the Doppler shift show that this luminous star is revolving about a mass center in a period of about 5.6 days. The spectrum of the other object does not show. This in itself is not remarkable, but type B stars are not known to be X-ray sources. However, the evidence is strong that the type B star is losing mass slowly and that this mass is streaming toward the invisible companion to form a disc from which the gas finally falls on to the invisible star. If the invisible object is a black hole, then even at its event horizon the pull of gravity is millions of times that at the surface of a white dwarf. The result is that before reaching the event horizon the gas, and particularly the electrons, are accelerated to velocities not far from that of light. As these high-speed electrons plunge into the accretion disc they collide with nuclei (mostly protons), are sharply slowed down, and lose their energy in the form of X-rays, which we can observe. The problem now is to decide whether the invisible object is a black hole or a neutron star; the latter might behave in the same way in producing X-rays. The solution—not an easy one—is to measure the mass of the invisible star. If it is two or three solar masses it is probably a neutron star, but if much larger it is probably a black hole. In the case of Cygnus X-1 one can make a fairly intelligent guess of the distance between the centers of the two components and that, together with the period and the presumed orbital inclination, when inserted into Kepler's third law gives the mass sum. One now needs to make an intelligent guess as to the B star mass and the remainder will be that of the invisible star. In spite of the assumptions, the evidence now is that the invisible object in this binary is probably a black hole, but it is not certain.

There is another system like the one above in which during the orbital revolution the X-rays are regularly cut off for a few hours and then reappear. This may be an eclipsing binary, which means that some restrictions are placed on the orbital inclination.

This picture of the production of X-rays by a black hole may well explain most X-ray sources, but not all. For example, the pulsar (neutron star) in the Crab Nebula also emits X-rays. At this time no one has suggested a means by which a black hole could also be a pulsar. As will be seen in the next chapter, black holes may well account for a number of previously unexplained extragalactic phenomena. If one considers that a black hole captures matter, it means that a black hole could increase its mass a great deal if it were able to swallow stars, as it surely must. Rather surprisingly it has been found that five globular clusters are X-ray emitters. The mass density at the center of a globular cluster is certainly quite high, and if one of the central stars had evolved into a black hole, the latter would have plenty of matter to feed upon to produce X-rays in the manner described above.

7.18 STELLAR POPULATION TYPES

Since about 1945 it has become apparent that there are distinct differences between the proportions of the chemical elements in the stars. It is now known that most of the stars in the flattened disc of our galaxy are relatively rich in the heavier elements (beyond helium) as compared with the galactic halo stars, such as those in globular clusters, and numerous other types. Stars are still forming out of the dust and gas clouds in the spiral arms of the disc, but there is no such material in the globular clusters. Hence, no new stars are forming there. Star-forming material is largely confined to the spiral arms. Because the first objects formed in the development of our galaxy were the globular clusters and the original primordial material must have been almost entirely hydrogen, the globular clusters are the oldest objects in our galaxy and still consist mainly of hydrogen.

However, we occasionally see in the disc of a galaxy the explosion of a massive supernova. At least 50 percent of the supernova's mass may be ejected to be mixed with the interstellar medium, which will then become enriched by appreciable amounts of the heavier elements produced by nuclear reactions that had taken place earlier in the exploded star. A supernova is the only source of any of the heavy chemical elements. As a result, the heavy elements found in the sun and the earth must have been created by supernovae a long time before our solar system began to form out of an interstellar globule of dust and gas. Now, new stars forming out of this enriched medium will have a quite different initial chemical composition than those formed earlier from nearly pure hydrogen. Physical considerations too complex to be discussed here show that a star's evolution over the ages will depend considerably on the heavy element proportion of the material from which the star was formed. Such structural differences have been observed. For example, the classical cepheids are Population I objects found largely in the spiral arms and with ages of just a few billion years. In contrast, the globular cluster cepheids and the Population II cepheids are found mainly in the galactic halo, where we see mainly the oldest stars that were formed out of nearly pure hydrogen. These objects are rarely found in the disc. Our sun is a Population I object 4.6 billion years old near the galactic plane and formed out of interstellar hydrogen clouds already enriched to some extent by the heavier nuclear debris from previous supernova explosions. The globular clusters have an age between 2 and 3 times that of our sun. Classification into just two population categories is an obvious oversimplification, and there is known to be a range all the way from stars with a small proportion of heavy elements to stars with a high proportion.

KEY TERMS

center of mass	classical cepheids	halo
spectroscopic binary	cluster cepheids	corona
eclipsing binary	pulsar	interstellar molecules

galactic clusters   Milky Way   white dwarf

globular clusters   disc   planetary nebula

cepheid variables   central bulge   black hole

novae

QUESTIONS

1. What is the sum of the masses of the two components of a double star if the period is two centuries and the separation is 7 billion miles?

2. What effects would you expect to observe in the light curve of an eclipsing binary if one of the stars had a large sunspot like those on our sun?

3. What might be expected to happen in our solar system if our sun became an ordinary nova? A supernova?

4. What changes would be observed in the light curve of an eclipsing binary if the stars were so close that the two surfaces facing each other were made hotter than the opposite surfaces?

5. Suppose that a globular cluster has a diameter of 20 light years and contains 100,000 stars whose average diameter is that of our sun. What fraction of the projected area of the cluster would be covered by the discs of the stars? Note that the chance that one star in the cluster would eclipse another would be small.

6. What is the difference in absolute magnitude of two stars of types B and K when both are on the main sequence?

7. What is the absolute magnitude of a Population I cepheid if it has a pulsation period of 10 days?

8. None of the stars in the sky that are visible to the unaided eye are type M main sequence stars. Suggest a reason for this.

8 Galaxies and Cosmology

Beginning with the use of large telescopes about 200 years ago and the advent of photography about a century ago, astronomers have discovered a large number of nebulous objects *not* confined to the Milky Way region, as are those bright nebulae that we now know are associated with the interstellar clouds. Photography has revealed these nebulous objects all over the sky. Most of them show a rather symmetrical structure, in contrast to the bright nebulae in the Milky Way, which are irregular in outline. Many of these symmetrical objects show a bright nucleus from which emerge spiral arms. Others look like bright globules that are circular or elliptical in outline. When counts were made of the number of these objects per square degree down to a given limiting magnitude, almost none were found in the Milky Way. Outside the Milky Way, the number of nebulae per square degree increases rapidly, going on to fainter magnitude limits. Good photographs of the spiral nebulae show that they are probably flattened, disc-shaped objects with a random spatial orientation, so they are observed on edge at one extreme and full-face at the other.

Among these spiral nebulae, the Great nebula in the constellation Andromeda has the greatest angular diameter. (Figure 8.1). This elliptically shaped spiral has an angular diameter of five degrees along its major axis (ten times that of our moon). The bright, central nuclear region is just visible to the naked eye; this spiral nebula is the only one that can be seen without a telescope. The outer spiral region has a surface brightness much less than that of the central nucleus and can be seen only on long-exposure photographs.

8.1 THE DISTANCE OF THE ANDROMEDA GALAXY

These spiral nebulae were first thought to be objects in our own galaxy, and many astronomers believed them to be rapidly rotating gaseous masses. However, about 1920, Mount Wilson Observatory astronomers resolved the outer portions of the Andromeda nebula into stars and showed that this object is really an assemblage, or kind of cluster, of stars. In 1924, Edwin Hubble of that observatory identified a few of these stars as classical cepheids with a mean apparent magnitude of about +18.

Continued observations accumulated enough data to obtain the period of the light variation. By means of the period-luminosity relation discussed in Chapter 6, Hubble was able to calculate for the first time the distance of a spiral nebula. The modern value for the distance of the Andromeda galaxy is about 2.2 million light years; therefore it is well outside the confines of our galaxy. Hubble determined distances for a number of the brighter galaxies and found that the Andromeda galaxy is the nearest spiral.

This galaxy is nearly the twin of our own—it has about the same linear diameter, mass, and luminosity. Like our own, it has spiral arms and interstellar clouds. Careful studies reveal that the resolved stars in the outer spiral arms are very bright blue stars of Population I, like those in our own spiral arms. It is possible to observe Milky-Way-type clusters in the spiral arms and to see that the Andromeda galaxy has a halo of globular clusters like our own. Occasionally, ordinary novae are seen. Radial velocity studies show that this galaxy is in rotation. The nuclei of both galaxies consist mainly of red, Population II giants and subgiants.

8.2 THE CLASSIFICATION OF GALAXIES

It has been clear for some time that the elliptical and spiral nebulae are really **galaxies** in their own right. They are, in fact, gigantic assemblages of stars. At one time it was fashionable to call them island universes. Of all these galaxies, the Andromeda galaxy is the nearest. However, our own galaxy has two companions called the Magellanic Clouds, which are classified as galaxies but are much smaller than our own. Each of these dwarf galaxies is about 200,000 light years from the center of our galaxy. They

FIGURE 8.1 Great galaxy in the constellation Andromeda. Star images over the whole area are foregound stars in our galaxy. (Palomar Observatory photograph.)

are seen close to the south celestial pole, and hence they are not seen in middle northern latitudes.

Galaxies may be classed by their appearance into three groups—spiral, elliptical, and irregular. About 75 percent are spirals, 21 percent ellipticals, and the remainder irregulars. Later on in this chapter, we shall see that this classification is too simple and needs modification.

The **spiral** galaxies have a nucleus from which the spiral arms (usually two) emerge. When the nucleus is relatively large, the arms are tightly wound around the nucleus; but if the nucleus is comparatively small, the arms are spread out and rather diffuse. Bright and dark nebulae are found in the arms together with blue giants of Population I. No doubt all the other stars on the main sequence are present, but they are too faint to be seen. With our present telescopic equipment, a star like our sun would be too faint to be resolved in the Andromeda galaxy. A typical view of a spiral galaxy seen full-face is that in Figure 8.2, which shows the nucleus and ragged spiral arms. A spiral seen on edge is beautifully shown in Figure 8.3. Note the pronounced central bulge of the nucleus and the dark obscuring clouds.

A subclass of the spirals is called the *barred* spiral (not shown). It consists of a luminous bar of stars which passes through the nucleus. From each end of the bar, a spiral arm begins.

The **elliptical** galaxies are elliptical or spherical in outline, as shown in Figure 8.4. They have no spiral arms and only rarely can one observe any obscuring clouds.

FIGURE 8.2 Spiral galaxy NGC 628, in the constellation Pisces, seen face on. Note the bright central nucleus and well-defined but somewhat ragged spiral arms. (Palomar Observatory photograph.)

THE CLASSIFICATION OF GALAXIES

FIGURE 8.3 The spiral galaxy NGC 4565 in the constellation Coma Berenices seen edge on. Note the spheroidal central nucleus and absorbing band of interstellar clouds. (Palomar Observatory photograph.)

FIGURE 8.4 Giant elliptical galaxy in the constellation Virgo. Observe absence of obscuring clouds and numerous globular clusters which appear as faint, fuzzy dots in the halo. (Palomar Observatory photograph.)

Recent studies show that they contain no appreciable amount of interstellar hydrogen. From this one can probably conclude that stars are no longer being born, and indeed the brighter stars are of Population II. In these respects, the elliptical galaxies resemble the nuclei of most spirals, whose brightest stars are of the same population. Some of the nearby ellipticals exhibit a halo of globular clusters.

The least numerous class of galaxies is the **irregular** type, which shows no symmetry of form or plan. Most irregulars are of low luminosity and are conspicuous for the presence of gas, dust, and Population I stars.

8.3 THE DETERMINATION OF THE DISTANCES TO GALAXIES

The use of the classical cepheids to determine the distances to galaxies applies only to the nearer ones in which the cepheids are bright enough to be resolved as individual stars (see Section 8.1). Beyond this, the problem becomes very difficult and less direct.

One method that applies fairly well to galaxies near enough to be resolved into stars is to determine the apparent magnitude of the brightest blue supergiant stars and to assume that their mean absolute magnitude is the same as that of such objects in our own galaxy, for which we have information on absolute magnitude. The same technique applies to galaxies whose globular clusters can be observed. Similarly, the astronomer can study the apparent magnitudes of a galaxy's ordinary novae and once again assume that their mean absolute magnitude is the same as for those in our galaxy. The difference between the mean apparent magnitude and the mean absolute magnitude of each class of object gives the distance modulus and hence the distance. The agreement for these classes of objects is good.

However, when it is impossible to resolve these classes of objects in a more distant galaxy, the methods of measurement become more difficult and uncertain. A technique that can be used is to make an estimate of the absolute magnitude of the whole galaxy and obtain the distance modulus from direct measurement of the apparent magnitude. The absolute magnitudes of nearby galaxies can be obtained with some precision because their distances are well known by other methods. The difficulty lies in the wide dispersion of the absolute magnitudes of galaxies taken without regard to class. The dispersion is much less if one regards only the larger spirals and uses them as standards. However, the image of a faint, distant galaxy is so diffuse that it may not be possible to tell to which class of galaxy it belongs. The difficulty may have been somewhat overstated here, but suffice it to say that methods have been derived by those who work in this field and that they give reasonably accurate distances. There are still many pitfalls, and the astronomers concerned are far from being satisfied with the present methods for very distant galaxies. As we shall see later, the problem of the structure and origin of the universe may well hinge on the accurate measurement of the distances of the most remote galaxies, just where the distances become so uncertain.

8.4 THE ABSOLUTE MAGNITUDES, DIMENSIONS, AND MASSES OF GALAXIES

From the apparent magnitude and the distance of a galaxy, one can determine its absolute magnitude. It appears that on the whole the spirals, with a mean M of -21, are the brightest galaxies. A few of the ellipticals are brighter, but their mean M is a little less. At the other end of the brightness scale are some of the dwarf elliptical galaxies, with M around -10, which is about equivalent to that of the brightest blue type O giants in our galaxy.

Again, the spirals are the largest galaxies, with diameters between 100,000 and 125,000 light years. Some of the dwarf ellipticals are no more than about 3000 light years across. Our knowledge of the number of dwarf galaxies in space is poor because they are so faint that only those nearby can be observed. The number of stars in an individual galaxy may range enormously from as few as about a million in a dwarf to perhaps a trillion or more in the larger spirals and elliptical galaxies.

The masses of galaxies are hard to determine, but considerable success has been achieved. In just the last few years our knowledge of galactic masses has changed a great deal, in part because of the use of larger telescopes and much more efficient devices for obtaining the spectra of very faint objects. Basically, the method depends upon the use of Kepler's third law in the form $a^3/P^2 = m_1 + m_2$ (Section 1.14). For our own galaxy this has been discussed in Section 7.11. In Section 1.14, where one determines the mass of a planet from the distance and orbital velocity of a satellite, it is assumed that the motion of the satellite is governed only by the planet's mass and that, if other satellites are present, their gravitational pull is negligible compared to that of the planet on the designated satellite.

For the mass determination of spiral galaxies one selects those seen approximately on edge but not too nearly edge-on. Thus the view of what is a galaxy with an approximately circular outline seen face-on now appears elliptical. Now place the slit of the spectrograph along the major axis of the elliptical image, obtain the spectrum, and measure the Doppler shifts *with respect to* the galaxy's *center*, of as many spectral lines as possible in the galaxy's spectrum. These Doppler shifts when converted to radial velocities will become the circular *orbital* velocities of the luminous matter orbiting around the galaxy's center. There will be a small but constant percentage correction necessary as a result of the inclination of the plane of that galaxy with respect to our line of sight. Knowing the galaxy's linear distance and the angular distance from its center to each point on the slit, it is now possible to make a simple plot of how the orbital velocity of luminous matter in the galaxy changes with distance from the center in light years or astronomical units. What is seen from such a plot is that, starting at zero at the galaxy's center, the circular orbital velocity increases steadily in about a straight line through the nuclear bulge region, suggesting that this region is rotating as a solid. Beyond this point the curve soon levels off and continues at this nearly constant velocity well out beyond the bright nuclear region. This is surprising because if (as originally thought) most of the galaxy's mass was found in the central nuclear region, the orbital velocity of matter well outside the nucleus should decrease steadily with distance from the center. This is what is observed in our solar system where 99.8 percent of the system's mass is in the sun and only 0.2 percent in the planets and satellites. Here, the orbital velocities of the planets steadily *decrease* with *increasing* distance from the sun.

But for those galaxies for which measurements of orbital velocity have been made, the velocities do *not* decrease beyond the bright nuclear region. This evidence strongly suggests that in the outer, faint parts of the galaxy there is a large amount of essentially nonluminous matter which contributes scarcely anything to the total light of the galaxy. Possibly this matter consists of great numbers of low-mass and very low-luminosity stars. It now appears that these galaxies, including our own, may well have a mass as much as a trillion solar masses or more instead of previous estimates of one hundred billion—an increase by a factor of about ten—or even more. We shall see later that this conclusion is of great importance in the calculation of the average mass density of the universe.

8.5 CLUSTERS OF GALAXIES—THE LOCAL GROUP

The evidence is now strong that perhaps all galaxies may be members of clusters of galaxies. The larger clusters contain thousands of galaxies and tend to be very compact. Clusters with a small number of members (Figure 8.5) have a somewhat irregular appearance. There are quite a number of pairs of galaxies. They are so much more numerous than would be expected on the basis of a random distribution on the celestial sphere of single galaxies that one can only assume that they are real physical pairs, in the sense that they revolve about each other. The periods of revolution are so long that no change in radial velocity is observed in a lifetime, but for many of the pairs, the radial velocity is not the same for both galaxies of a pair. In this sense, the members of the pair behave like the components of a spectroscopic binary, and from this information it is possible to derive some information on the sum of the masses of the pair.

For a number of nearby clusters in which the galaxies are fairly bright, the radial velocities of a large number of cluster members have been measured. In any one cluster, there is a fairly large spread in the individual radial velocities of the different members. This range may be in the neighborhood of 1000 km/s (600 mi/s). This indicates that there are internal motions in the cluster; these motions are the result of

FIGURE 8.5 A portion of the galaxy cluster in the constellation Virgo. (Courtesy Kitt Peak National Observatory.)

the gravitational attraction of the members for each other. The more mass in a given volume, the greater the internal motions will be. The spread of these motions around the average for the cluster can be used to make a fairly reliable measure of the total mass of the whole cluster. However, when this mass is compared with that obtained by summing up the estimated masses of the individual members, it is found that the sum obtained from the radial velocity spread is in many cases anywhere from 10 to 50 times greater. The "missing mass" has been a difficult problem for astronomers, but part of this discrepancy may be accounted for by the newer results for individual galaxy mass measures as discussed in Section 8.4. There is now evidence that between the galaxies in a cluster there may be a large amount of highly ionized transparent gas (mostly hydrogen) which would help to account for the "missing mass" discrepancy.

Our galaxy, the Andromeda galaxy, and about 27 others are known to be members of what is called the **local cluster**, which has a diameter of not quite three million light years. Of these 29 galaxies, 3 are spirals, 14 are elliptical types of all sizes, and 12 are irregulars. Twelve of the ellipticals are classed as dwarfs. The two Magellanic Clouds mentioned earlier are irregular galaxies. Other dwarfs may be present in this cluster, but they are probably too faint to be distinguished against the foreground stars in our galaxy, particularly in the Milky Way region. The Small Magellanic Cloud now appears to be two dwarfs seen in nearly the same line of sight.

8.6 THE VELOCITY-DISTANCE RELATION—THE RED SHIFT

About 1915, astronomers began to obtain spectra of a number of the brighter galaxies within reach of existing telescopes. These spectra are the integrated spectra of a large number of stars in the galaxy. Because of the internal motions in the galaxy, most of the fainter stellar absorption lines are smeared out due to different Doppler shifts. However, some of the stronger lines show, for example, the H and K lines of singly ionized calcium at the respective wavelengths of 3969 Å and 3934 Å. Galaxies with bright interstellar clouds show, in addition, a number of emission lines coming mainly from gas clouds illuminated by hot, very bright stars.

Radial velocity measurements of these spectra show that, except for some galaxies in the local cluster, all of the spectra are shifted toward the *red* region of the spectrum. The most generally accepted explanation of these red shifts is that they result from the recessional velocities of the galaxies. It becomes apparent that the recessional velocities, on the whole, are greater for the fainter galaxies than for the brighter ones. Therefore, the recessional velocities increase with the distance of the galaxies. Figure 8.6 is a plot of the radial velocity against the distance for a large number of galaxy clusters. Although the radial velocities are considered to be quite reliable, there may be some serious uncertainties in the distances. These become greater for the more distant galaxies.

8.7 INTERPRETATION OF THE VELOCITY-DISTANCE RELATION

In Figure 8.6 it can be seen that the trend of velocity with distance can be represented rather well by a straight line. The slope of the best straight line through this scatter of points represents the rate at which the recessional velocity increases for, let us say,

FIGURE 8.6 The velocity-distance relation for clusters of galaxies. Each of the twenty data points is the average of the radial velocity and the distance for each of five galaxy clusters, arranged according to distance, for a total of 100 clusters. A best straight line is drawn through the first eighteen data points starting from the lower left. The nineteenth is close to this line but slightly above and the twentieth, the uppermost point, is well above the straight line, giving the first positive suggestion that the slope of the relation is increasing and that the recession rate was greater in past time. See Section 8.10. (These data are from the paper by J. Kristian, A. Sandage, and J. A. Westphal under the partial title "The Extension of the Hubble Diagram. II," which appeared in the *Astrophysical Journal*, Vol. 221 pp. 383–394, 1978.)

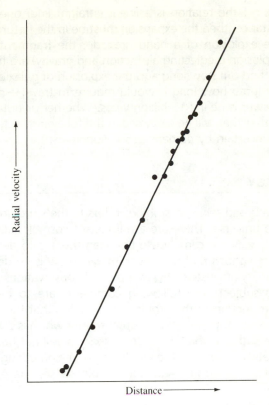

each million light years of distance increase from our galaxy. Here we shall adopt a value of 20 km/s (12.5 mi/s) per million light years. Because of uncertainties in the distance scale the actual value is probably somewhere between 15 and 30 km/s per million light years. This recessional rate is called the Hubble constant. Thus, on the average, the radial velocity of a galaxy at one million light years would be about 20 km/s. At 100 million light years the radial velocity of the galaxy would be about 100 times greater, or 2000 km/s (1250 mi/s). At any distance the radial velocity of a galaxy should be about equal to the Hubble constant multiplied by the distance in millions of light years.

That these galaxies are receding from us does not mean we are at the center of the universe of observable galaxies. Imagine a large sheet of rubber, which represents a two-dimensional universe, and let there be scattered on this sheet a number of white dots to represent galaxies. Let our galaxy be one of them. Now slowly stretch the sheet at the same constant rate in both dimensions. From our own dot, all the other dots will be seen to be receding from ours. The more distant the dot, the faster it will recede. This is exactly what is observed for galaxies. However, the same observations of radial motion from any other dot will show that all the others are receding from it, including our own. Transformed into three dimensions, this analogy will show that the universe will be observed to be expanding, no matter from what point it is observed.

If the relation is a linear (straight line) one between recessional velocity and distance, then the expansion must be in the nature of an explosion. At any time after the explosion of a hand grenade, the fragments most distant from the point of explosion (neglecting air friction and gravity) are those that have traveled the fastest. To find out how long ago the explosion of galaxies took place, it is only necessary to compute how long it would require to travel 1 million light years at 20 km/s. The answer is nearly 15 billion years. Whether or not this is the age of the universe is a matter that will be discussed in the following sections. At the present time this figure is uncertain by perhaps ±5 billion years.

8.8 COSMOLOGY

In a broad sense, *cosmology* refers to the structure, content, origin, and evolution of the universe. There are a number of cosmological models of the universe, and the only way we can choose between them is to decide which model agrees with the observations that have been made. There is no dearth of models, but there is a real scarcity of trustworthy, interpretable observations. Cosmology is a broad and intriguing subject. The following comments are to be considered only as a very brief introduction to the problem. It is hoped that the reader will consider the references given in Appendix 1. One other matter which is almost a matter of embarrassment to the author is that in this short space it will not be possible to consider the impact of relativity theory on the whole problem. Something will be lost in the telling, but it will not be disastrous. Again, if you please, see the Appendix for further reading.

8.9 THE BIG-BANG THEORY

The observable universe of galaxies appears to be expanding: no other explanation has been proposed that successfully explains the observations. If the velocity-distance relation is correct, it implies that all the matter of the universe was in the same place about 15 billion years ago and that at that time, for some unknown reason, it began to expand with great violence. If the explosion were of the hand grenade type and there were no gravitational retardation, the relation between the velocity and the distances of the fragments would be a linear one over all distances.

The big-bang theory assumes that at this very distant time past, all the matter of the universe was compressed into an enormously condensed state in a very small volume. The theory postulates that this primeval fireball consisted of a mixture of matter and radiation at a temperature of billions of degrees. As this universe began to expand, the temperature dropped rapidly, and in time the matter condensed into galaxies of gas and dust and finally into stars. The radiation density decreased, and along with that, the temperature of space decreased. Under these assumptions, all the galaxies are becoming more distant and fainter, and at some very distant future time, even the brightest galaxies will be too faint to be observed. In this theory, the universe is approaching an infinite extent. Perhaps the strongest argument in favor of the big-bang theory is the confirmed observation by radio astronomers of an influx of radiation which is constant in intensity from all directions in space and which has a spectral energy distribution corresponding to nearly 3K. This is thought to be the remnant of the original fireball radiation after some 15 billion years of expansion and cooling.

8.10 THE OSCILLATING UNIVERSE

Is there any alternative to the situation outlined above? The concept that the universe had a beginning and will expand for an infinitely great time is somewhat disturbing and it is clear that it has theological implications. There must exist a gravitational attraction between the galaxies and this should cause a slow decrease of the Hubble constant everywhere in the universe. If this does occur, one should observe that the velocity-distance relation is not a straight line, but is a gently sloping curve that changes upward from the straight line by an amount that differs more and more with increasing distance. The upward slope results from the travel time of light through space. The light we observe and the Doppler shifts we measure for galaxies 5 billion light years away left those galaxies around 5 billion years ago. If the Hubble constant is not constant but decreasing with time, then we ought to observe that the rate of velocity change with distance is greater for distant galaxies than that measured for relatively nearby ones. In other words, if one wants to study the behavior of the universe at an earlier time, the behavior of distant galaxies should be studied. The behavior in the relative present would be obtained from relatively nearby galaxies. This is in accord with the **cosmological principle**, which assumes that observations of the universe made from widely separated stations in the universe at the *same time* will always give the same results. However, the recently published data in Figure 8.6 do begin to show a slight upward trend away from a straight line through the data points. If this departure from a straight-line slope is real then this is evidence that the recessional rate was greater at some earlier time. What is needed are data for radial velocities at greater distances and a more reliable distance scale.

However, let us suppose that the velocity-distance relation did curve upward and that the recession rate is decreasing with time. This need not mean that the universe will stop expanding at some future time, because the present rate of expansion might be greater than the escape speed; this would produce the same result discussed in the last paragraph. To settle this problem, one needs to know the average density of the universe. To calculate this, one adds up the observable mass in all its forms in a large volume of the universe and divides by the volume. When this is done and the escape velocity is calculated, it turns out that the escape speed is much less than the observed expansion rate. A further calculation shows that in order to make the calculated escape speed greater than the observed expansion rate, it is necessary to find at least ten or more times the present observable mass. See Section 8.12 for the possibilities of doing so.

If further work shows that the expansion rate is actually decreasing with time and that at present it is less than the escape speed, then in time the universe should come to rest and begin a collapse. After a very long time, all the matter in the universe would smash together to form a fireball and then, perhaps, the expansion would begin all over again. Under these circumstances, the universe would be an **oscillating** one. Thus, the universe may have expanded and collapsed for an infinitely long time in the past and will continue to do so for an infinitely long time in the future. There would be no beginning and no end.

8.11 THE STEADY-STATE THEORY OF THE UNIVERSE

This theory states that the universe is infinitely old, accepts the observation that it is expanding, and assumes the cosmological principle. Therefore, the universe as it appears now has been the same for an infinitely long time in the past and will retain the same general appearance for all future time. It assumes that the space density of galaxies is always the same, regardless of expansion. To accomplish this, the theory suggests that new galaxies are constantly being created to replace those that are expanding out of sight. The theory proposes that matter is being continuously created at the proper rate per year to keep the space density constant. This violates the law of conservation of matter and energy, which is one of the cornerstones of modern physics, although no experiment can ever be devised to disprove the law. However, the amount that would need to be created per year is far too small to be detected in any laboratory.

The steady-state theory was much favored at one time, but it lost much ground after the discovery of the 3K radiation (see Section 8.9), which is not predicted by this theory.

8.12 INTERGALACTIC MATTER

In view of what has been said in Section 8.10, it is of great importance to arrive at a reasonably accurate value of the mass density of the universe. When this is done it should be possible to decide whether or not the universe is an *open* one (expanding indefinitely in time) or a *closed* one (oscillating model). Until recently the mass density was too small by a factor of at least ten to conclude that the universe will not expand indefinitely. Recent work has changed that picture somewhat in the direction of previously undiscovered mass.

Astronomers are not keen on finding a considerable amount of intergalactic dust because this dust, even though thinly distributed, would dim the galaxies and affect the calculated distances. Therefore, one might look for gas, which is far less effective in obscuration. If, as generally believed, the galaxies condensed out of gas at a very early stage of their evolution, there ought to be some uncondensed gas left, particularly in the clusters of galaxies and especially so in the richest clusters.

One crucial piece of evidence for intergalactic matter comes from the internal motions of the individual galaxies in a cluster. Just as a strong spring will oscillate more rapidly than a weak one, so should the internal motions be greater in a massive cluster with many galaxies than in another cluster of the same size with fewer galaxies. Therefore, it is reasonable to expect that a measurement of the radial velocity spread of the individual cluster members will yield a determination of the cluster mass. It is necessary to assume that the clusters themselves are not expanding, and it is generally believed that they are stable. The measured mass so determined is much greater than the mass estimated from the counts of galaxies in the cluster— greater by a factor of about ten. The missing mass could be accounted for partly by dwarf galaxies too faint to be seen, by stars that had escaped from galaxies, and by intergalactic gas within the cluster.

There is now evidence from radio observations and other sources that many galaxies are surrounded by a halo of hydrogen and that, in some cases, the hydrogen may increase the mass of the galaxy by a factor of ten. For example, the giant elliptical galaxy M87 emits strongly in the radio and X-ray region. Part of the radio noise may come from the halo hydrogen. For this galaxy it can be shown that there is a central *nonluminous* mass concentration of about five billion solar masses—possibly a gigantic black hole. This also brings up the possibility that there might be a host of black holes of more modest masses that began their life as massive stars.

In the total of the mass of any given large volume of space one must include the equivalent mass of the optical light and all other forms of radiation. It begins to appear that the use of modern techniques and new discoveries is beginning to reduce the amount of missing mass, thereby increasing the average mass density of the universe. Whether or not we can account for enough to stop the present expansion at some distant time is open to question, but it is not as uncertain as some thought ten years ago.

8.13 / EXTRAGALACTIC RADIO SOURCES

Although the great majority of galaxies appear to be radio quiet, there are some that emit enormous quantities of radio energy. The Andromeda galaxy and our own galaxy are weak emitters of radio energy and are regarded as radio quiet. The galaxy Centaurus A in Figure 8.7 is a powerful radio source. The photograph suggests that it is a globular galaxy, but the dark band of dust is almost unknown in this type. The photograph was taken, of course, in light from the optical region, but a radio telescope will show that the radio noise (synchrotron radiation) comes not from the center of the galaxy but from all over the region shown and from well outside, particularly strongly from the regions within the circled areas. With the increased resolution of radio telescopes in recent years, it has been shown that most **radio galaxies** are double sources of radio noise rather symmetrically placed on opposite sides of the optical image and at about the same distance. It appears that for some reason an explosion of great violence took place at the galactic center and ejected

FIGURE 8.7 The radio galaxy Centaurus A. While the radio energy comes from all points in the photograph and well outside, the strongest sources are within the two circles. (Palomar Observatory photograph.)

two masses of gas in opposite directions at nearly right angles to the plane of the galaxy. The two gaseous masses are moving outward at great speed and emitting synchrotron radiation. In many cases there is evidence that more than one explosion, separated in time by millions of years, has taken place. A sizeable fraction of the radio galaxies show a quite different pattern in the distribution of the radio noise sources. However, all appear to have in common a violent explosion, probably in the nucleus.

The total situation regarding radio galaxies is far more complicated than that indicated in the preceding paragraph. Seyfert galaxies are spirals with exceptionally bright nuclei. Spectra of the nuclei show broad emission lines indicative of large internal motions. BL Lacertae objects are very strong radio emitters, and in the optical region they are starlike on a photographic plate. A few show wisps of nebulosity, suggesting that they may be the nuclei of galaxies. These same objects show very faint absorption lines on a continuous background, and measures of the lines show velocities of recession. There are other types of radio emitters which may be galaxies, but the situation as of now is most confusing, particularly if one is looking for an evolutionary sequence. One of the most interesting types is the **quasar**, which is discussed in the next section.

8.14 QUASARS

As radio telescopes became more sensitive and their resolution improved, it became clear about 1960 that the sky contained a large number of "point" radio sources, that is, radio sources of small angular diameter. Some of these sources were identified with starlike images in the optical region, and they did not appear to be galaxies. All of the original objects were intensely blue. Optical spectra showed a number of emission lines whose wavelengths did not correspond with the usual elements found in hot stars. The puzzle was solved in 1963, when it was found that the disparity could be explained by the fact that the usual spectral lines were all red shifted by the same fraction, $\Delta\lambda/\lambda$. An object of this type has been named a **quasar** for quasi-stellar object.

From the start it was generally assumed that these red shifts were Doppler shifts caused by radial velocity. In Section 3.14, the expression for radial velocity is given by $v = \Delta\lambda/\lambda \times c$, which can also be written as $v = z \times c$, where $z = \Delta\lambda/\lambda$. For the first quasar discovered, $z = 0.16$. This is an enormous value. Inserting this value of z in the relativistic formula in Appendix 2, III gives a radial velocity of 44,200 km/s (27,500 mi/s). It should be emphasized here that for *all* quasars the shift is to *longer* wavelengths; and if these are Doppler shifts, all the velocities are recessional—*not one* is blue shifted. In later work some values of z were found to be greater than unity, which would indicate from the formula that their velocities were greater than that of light. But, if the theory of relativity is correct, no body can move faster than light. However, the theory does provide a formula given in Appendix 2, III for any value of z. For the reader with a little algebraic background, it is easy to show that no matter how large the value of z, the velocity can never be greater than that of light. At the time of this writing, the largest value of z is 3.78. Again, inserting this value of z in the formula in Appendix 2, III shows that this quasar is receding at 91.6 percent of the speed of light or very nearly 275,000 km/s. Assuming that the Hubble constant of 20 km/s per million light years is constant to all distances, this number divided into

275,000 km/s gives the quasar's distance as close to 13.8 billion light years. Therefore, we are seeing what that quasar was doing 13.8 billion years ago.

If the red shifts of the quasars are cosmological in origin (due to the expansion of the universe), then they are more distant than any galaxy that can be recognized as such. Even though their apparent magnitudes are rather faint, the quasars must be enormously bright objects—brighter by something like a million times than the brightest ordinary giant galaxies. If they really have these enormous intrinsic brightnesses, then what process is in operation that would produce this enormous outpouring of energy? Nuclear processes, supernova explosions, and collisions between stars, either separately or together, cannot supply the energy. However, the most recent suggestion is that a quasar is the small nucleus of a galaxy. This nucleus contains a large black hole, into which stars are falling. Before the stars had reached the event horizon the gravitational field of the black hole would disintegrate the stars with the release of large amounts of energy. Clearly this process cannot go on indefinitely because in time the matter source will be exhausted. Although the black hole could contain most of the matter of the galaxy, there could remain enough unconsumed matter well outside the event horizon to produce the observed spectral lines. Some quasars show double or triple absorption lines and some even have a few emission lines, which only complicates the model. At one time it was thought that all quasars were radio emitters, but it is now known that the majority are not and that most are radio quiet. And to complicate matters further, a few of the quasars with large z values are not intensely blue. Obviously, the nature of quasars is very intriguing.

Some astronomers prefer to believe that quasars are not really that distant. They believe those we observe were ejected from the nucleus of our galaxy with enormous speeds at a time long enough ago for those coming toward our position to have passed us; so all have recessional velocities. If this were true, then one should expect that this should have happened in other nearby galaxies, giving some of the quasars approach velocities and blue shifts. However, *all* wavelength shifts are to the red. A more probable possibility is that at least part of the red shift is gravitational. Relativity theory shows that the light escaping from the surface of a luminous body should have its wavelengths made longer as a result of having escaped from the gravitational attraction of the object, and the amount of reddening would be proportional to the surface gravity. This could be at least a partial explanation of the quasar red shifts if they were highly condensed objects of large mass. The general weight of evidence is that the red shifts are cosmological, and it is nearly conclusive.

Another significant observation on quasars is that some of them fluctuate in brightness more or less randomly in times ranging from about a day to years. This tells us something about their size. Consider a distant supergiant star whose radius was one astronomical unit, or about 8 light minutes. If this object would suddenly increase its surface brightness by, say, 10 percent, we would first observe the increased light from the center of the surface closest to us and then the light from those areas farther from us, and finally the increased light from the edge. The result would not be a sudden increase of perhaps a few seconds but a slow increase lasting for 8 minutes. Therefore, a quasar which exhibits a light increase lasting for about a light week would have a radius of about a light week, which is about 1/50 of a light year. There is now some evidence that quasars may be galaxies in which the energy production is taking place in a small nucleus or in a quite small part of a much larger nucleus where the mass density is very high.

If it should turn out that quasars are very distant objects, then for many of them their light has been a long time on the way. We would be seeing in them what the universe was doing many billions of years ago.

8.15 THE ORIGIN OF GALAXIES

Some astronomers believe that what we are *now* observing in the quasars is galaxies in formation, and that the radio quasars are either a short-lived phase or one that repeats itself at intervals in the early stage of the formation of most, if not all, galaxies. It has been proposed that not long after the initial big-bang explosion, the matter (mostly hydrogen) fragmented into very large masses far greater than that of the average galaxy. As these masses began to contract because of their own gravitation, they also started to fragment into proto-galaxies, and soon individual stars began to form. There appears to be an upper limit to the mass of a galaxy. The result was the appearance of clusters of galaxies. Quite possibly, many of these galaxies went through an unstable quasar stage early in their history and finally settled down to the kind of well-behaved galaxies that we now see in our neighborhood. If the above is a reasonable approximation, then the quasars should all be distant objects.

Again, if one can accept the above picture, then this is one more reason to reject the steady-state theory. This theory argues for the continuous creation of galaxies, which ought to mean that we should observe some quite near quasars; we do not.

8.16 SOME UNSOLVED PROBLEMS IN GALACTIC EVOLUTION

Most of the statements in the above section sound fairly reasonable, but there are a number of loose ends. Color measurements show that the irregular galaxies are predominantly blue, the ellipticals reddish, and the spirals somewhere in between. In Chapter 6, it was stated that a blue star is very hot, very young, and evolving very rapidly into the red giant stage, after which it will begin its collapse to the white dwarf stage or some other type of supercondensed star. The red main-sequence dwarfs are too faint to be seen in any galaxy but our own. On the basis of this evolutionary behavior, it would be assumed that the irregular galaxies are young, the spirals somewhat older, and the ellipticals the oldest. But because the blue supergiants of an irregular galaxy burn out rather rapidly, they must have been created a short time ago. Could they have come from the ellipticals and, if so, by what process? Only rarely do ellipticals show any interstellar material in the galaxy; the irregulars would have to be produced from something else. It has been known for some time that the elliptical galaxies are in some cases more massive than the spirals by perhaps as much as 50 times. If these massive ellipticals are the parents of the spiral galaxies and others, then why do the spirals have arms with interstellar clouds and blue giants, while the elliptical galaxies do not? It is of interest to note here that the nuclei of most spirals resemble elliptical galaxies.

The situation is in a not-very-satisfactory state, but that is the usual condition in a rapidly developing field. In fact, it is this unresolved aspect of problems which has great appeal to scientists. A mass of observational data, plus a number of conflicting theories, present to scientists a great challenge and one that they may well face with considerable enthusiasm. They also know that in the solving of one problem they will uncover others that will challenge their attention and interest.

KEY TERMS

spiral nebulae	our local cluster	oscillating universe
spiral galaxy	Magellanic Clouds	cosmological principle
elliptical galaxy	velocity-distance relation	radio galaxies
irregular galaxy	Hubble constant	quasar
clusters of galaxies		

QUESTIONS

1. Summarize the several methods for the determination of the distances of galaxies. What would be the parallax in seconds of arc of the Andromeda galaxy?

2. Discuss the presence of interstellar material in the classes of galaxies and the consequences for the evolution of galaxies.

3. What is the evidence that galaxies are in rotation?

4. What might you expect to observe during and after the collision of two galaxies?

5. What difficulties are encountered in discovering new dwarf galaxies?

1 Appendix

SUGGESTED READINGS

The following textbooks are recommended for in-depth reading:

Abell, George. *Realm of the Universe*, 3rd ed. Philadelphia: Saunders College Publishing Co., 1984.

Beatty, J. K., O'Leary, Brian, and Chaikin, Andrew. *The New Solar System*. Cambridge, Mass.: Sky Publishing Corp., 1982.

Bok, Bart J., and Bok, Priscilla F. *The Milky Way*, 5th ed. Cambridge, Mass.: Harvard University Press, 1981.

Hartmann, William K. *Moons and Planets*, 2nd ed. Belmont, Calif.: Wadsworth Publishing Co., 1983.

Mitton, Simon, Editor-in-Chief. *The Cambridge Encyclopedia of Astronomy*. New York: Crown Publishers, Inc., 1980.

Pasachoff, Jay M., and Kutner, Mark L. *University Astronomy*. Philadelphia: W. B. Saunders Co., 1978.

Snow, Theodore P. *The Dynamic Universe*. St. Paul, Minn.: West Publishing Co., 1983.

In addition to the above, the articles which appear frequently in the monthly magazine, *Scientific American*, are highly recommended for their clarity and timeliness. A list of reprints on astronomy and other topics from this magazine may be obtained from W. H. Freeman and Company, 660 Market St., San Francisco, California 94104.

The monthly publication, *Sky and Telescope*, available in most libraries, is written particularly for amateur astronomers. It is well written and highly informative, and because of its up-do-date coverage of astronomical discoveries, it is read by many professional astronomers. The publisher of this magazine is Sky Publishing Corporation, 49 Bay State Road, Cambridge, Massachusetts 02138; from this firm one may obtain free on request their publication catalogue entitled *Scanning the Skies*. This catalogue contains their large offerings of star charts of several kinds, books on many aspects of astronomy, moon charts, books on how to make a telescope, and a wide selection of photographic prints of astronomical instruments and celestial objects. For general-purpose constellation study, *Norton's Star Atlas and Reference Handbook* (17th ed. Cambridge, Mass.: Sky Publishing Co., 1978) is highly recommended.

Catalogues of lantern slides and prints in both black and white and color may be obtained by writing to:

Hansen Planetarium
15 South State St.
Salt Lake City, Utah 84111

Hansen Planetarium is supplier of these materials from

a. Mount Wilson and Las Campanas Observatories
b. California Institute of Technology, Astronomy/Palomar Office
c. Kitt Peak National Observatory
d. Lick Observatory, University of California

A catalogue of these materials from an independent source may be obtained by writing to:

Yerkes Observatory
Photographic Services
Williams Bay, Wisconsin 53191

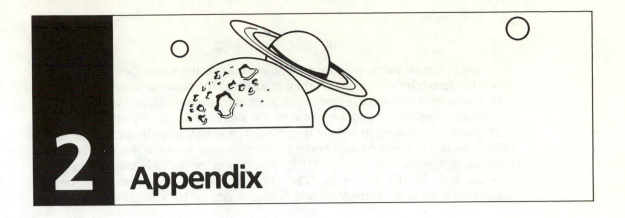

2 Appendix

I. METRIC AND ENGLISH UNITS OF LENGTH AND MASS

In contrast to the English system of length, all units in the metric system are related by powers of ten. Originally, in the metric system, the meter was intended to be one ten-millionth of the distance between the equator and the Pole. Below are given some conversions for length and mass between the two systems and their abbreviations.

Length

1 km = 1 kilometer = 1000 meters = 0.621 mile
1 m = 1 meter = 100 centimeters = 39.37 inches = 3.281 feet
1 cm = 1 centimeter = 10 millimeters = 0.3937 inch
1 mm = 1 millimeter = 0.1 centimeter = 0.001 meter

1 mile = 1.609 km, 1 foot = 30.48 cm, 1 inch = 2.54 cm (exactly, by international agreement)

Mass

1 kg = 1 kilogram = 1000 grams = 2.205 pounds
1 g = 1 gram = 0.0353 ounce
1 lb = 453.6 g. 1 oz = 28.35 g

II. TEMPERATURE SCALES

On the Fahrenheit scale, water freezes at 32°F and boils at sea level pressures at 212°F. The scale used in much scientific work is the Centigrade scale, in which the corresponding temperatures are 0°C and 100°C. Thus, between the freezing and boiling points of water there are 180 Fahrenheit divisions and 100 Centigrade divisions. To convert Fahrenheit to Centigrade temperatures, first subtract 32°F and multiply the result by $\frac{5}{9}$ (100/180) as in the formula

$$°C = \frac{5}{9} \times (°F - 32)$$

or

$$°F = (9/5 \times °C) + 32$$

Another scale much used by physical scientists is the Kelvin (absolute) scale in which the zero point is that temperature at which all molecular activity ceases (which is also the lowest attainable temperature). Absolute zero is nearly $-459°F$ ($-273°C$). To convert from the Centigrade scale to the Kelvin scale, simply add 273 to the Centigrade temperature. It is now becoming increasingly popular to use the name Celsius instead of Centigrade in honor of the man who invented this scale. Furthermore, a temperature of, say, $-112°C$ would now be written longhand on the absolute scale as 161 kelvins ($+273 - 112$) (lower case k), without the degree superscript used in the Fahrenheit and Celsius scales, or 161K in abbreviated form.

III. RELATIVISTIC FORMULA FOR RADIAL VELOCITY

When the radial velocity is small (say, less than 1.0 percent of the speed of light), the formula $v = \frac{\Delta\lambda}{\lambda} \times c$ is perfectly adequate for the determination (see Section 3.14). However, when $z = \frac{\Delta\lambda}{\lambda}$ is somewhat greater than about one percent, one should use the following relativistic formula:

$$v = c\left[\frac{z^2 + 2z}{z^2 + 2z + 2}\right]$$

For example, if $z = 0.05$ (5 percent), the first formula gives $v = 14{,}990$ km/s while the second (correct) relativistic formula gives 14,615 km/s. The difference is 375 km/s, an error of about 2.5 percent. However, since most radial velocities (such as those found for spectroscopic binaries) rarely are larger than 200 km/s the error in using the first, simpler formula is negligible. But consider for a moment that $z = 1$. The first formula would say that the radial velocity was equal to the speed of light and that any value of z greater than one (1) would give velocities greater than that for light, which is contrary to relativistic theory. When $z = 1$ is substituted into the correct relativistic formula, the radial velocity is only 179,875 km/s, a value much less than the speed of light: 299,792 km/s (186,282 mi/s). In the above relativistic formula for the radial velocity one can see that no matter how large z becomes, the radial velocity will always be less than the speed of light.

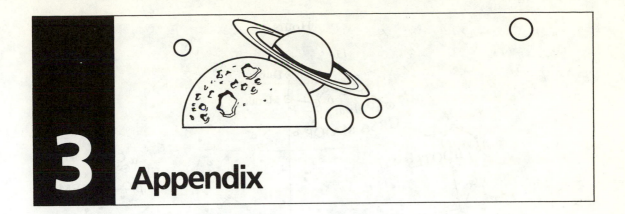

3 Appendix

STAR MAPS

The five star maps are charts of the brighter stars that can be seen in the sky by an observer in the Northern Hemisphere. On the four seasonal charts, right ascension is given along the top, and declination on the left side. The months along the bottom of each chart refer to the stars that will be (approximately) on the observer's local meridian at about 8 P.M. standard time. As the night progresses, the observer's local meridian moves from right to left across each chart. The declination of those stars that pass through the observer's zenith (overhead point) will be equal to the observer's latitude. On the circumpolar chart note that the pointer stars, a and β, of the Big Dipper (Ursa Major) point directly to the North Star (Polaris), which is a part of the Little Dipper (Ursa Minor).

The star maps have been reproduced (by permission) from *The Elements and Structure of the Physical Sciences* by J. A. Ripley, Jr., published by John Wiley & Sons, Inc. (1964).

159

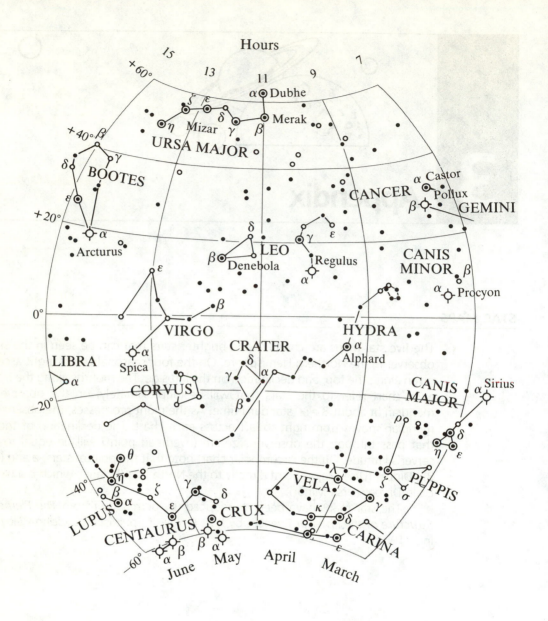

Hours

FIGURE A3.1 The spring constellations.

<svg width="20" height="20"><circle cx="10" cy="10" r="4" fill="none" stroke="black"/><line x1="10" y1="0" x2="10" y2="6"/><line x1="10" y1="14" x2="10" y2="20"/></svg> Brighter than magnitude 1.5

◉ Between magnitudes 1.5 and 2.5

○ Between magnitudes 2.5 and 3.5

• Fainter than magnitude 3.5

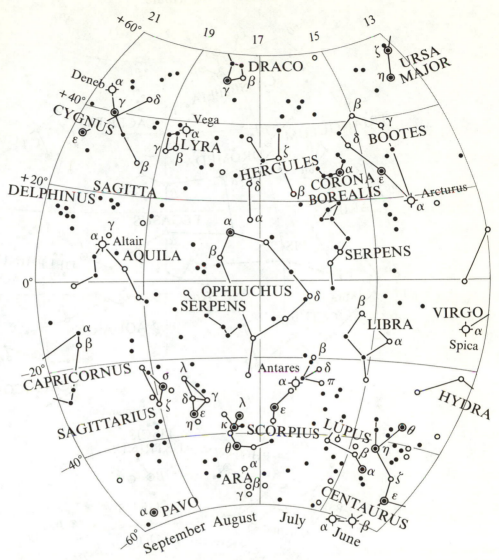

FIGURE A3.2 The summer constellations.

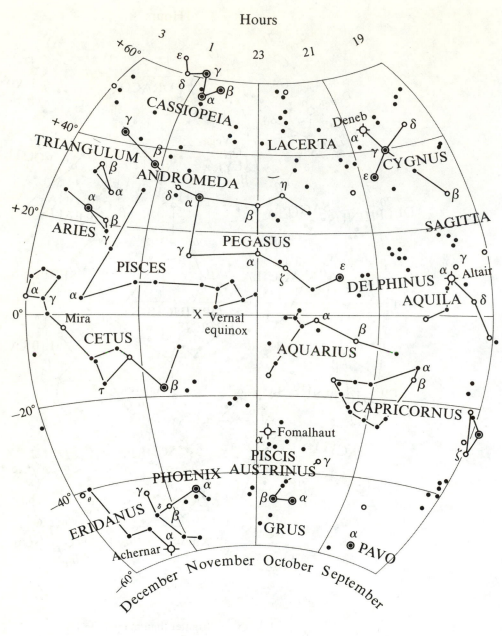

FIGURE A3.3 The autumn constellations.

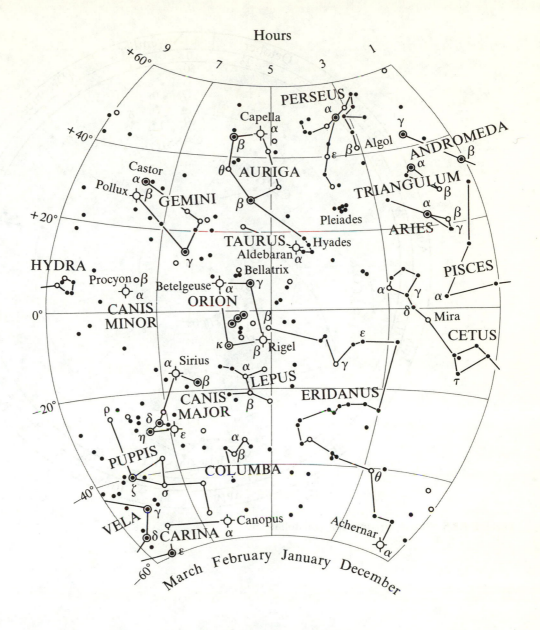

Hours

FIGURE A3.4 The winter constellations.

◇   Brighter than magnitude 1.5

◉   Between magnitudes 1.5 and 2.5

○   Between magnitudes 2.5 and 3.5

•   Fainter than magnitude 3.5

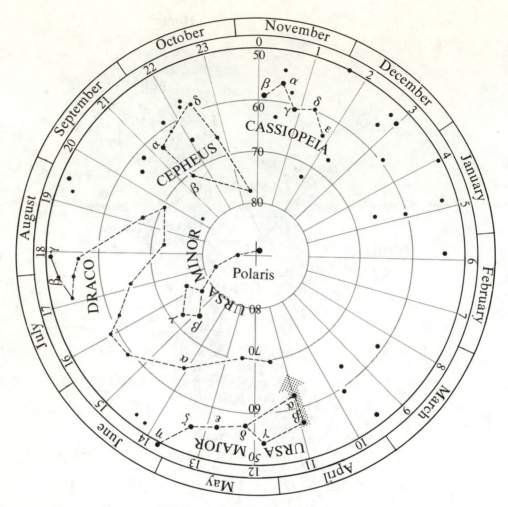

FIGURE A3.5 North circumpolar constellations.

Index